W9-CLQ-101

Networks in Aviation

Philipp Goedeking

Networks in Aviation

Strategies and Structures

Dr. Philipp Goedeking
airconomy aviation intelligence GmbH & Co. KG
Frankfurt Airport Center 1
Hugo-Eckener-Ring
60549 Frankfurt
Germany
e-mail: goedeking@airconomy.com

ISBN 978-3-642-13763-1 e-ISBN 978-3-642-13764-8

DOI 10.1007/978-3-642-13764-8

Springer Heidelberg Dordrecht London New York

Library of Congress Control Number: 2010934781

Cover design: WMXDesign GmbH

Printed on acid-free paper

Springer is part of Springer Science+Business Media (www.springer.com)

Foreword

Networks are everywhere. Each day, we deal with an array of social networks: the Internet, company intranets, electronic circuitries, chemistry, animal behavior, mathematics, textile fabrics, and tightly networked banks. In the field of aviation, networks abound: airlines offer networks of flights; alliance systems include networks of airlines; airports represent a network of global infrastructure; and air traffic control centers ensure safe distance between individual aircraft.[a]

Ever since the beginning of civil aviation, the structures and strategies of aviation networks have undergone fundamental change. Few if any other major global industries are subject to as many cyclical ups and downs as the aviation industry. Airlines are subject to constant regulatory change—from traffic rights to national ownership restrictions to environmental issues. They also are challenged by fundamental changes in distribution technology, such as sharp increases in direct sales through the Internet or call centers and lower entry hurdles for new entrants like low-cost carriers.

Aviation networks exert an enormous leverage on capital and operational expenditures, on job creation and job elimination, and on the economic perspectives of entire regions. All these forces are constantly reshaping the success factors of aviation networks. By the time an airline has developed and implemented a new network strategy and corresponding structure, the market may already have moved away. All facets of network economics are under constant pressure, and are viewed as untapped opportunities to increase revenues and lower costs.

Network strategies and structures are the result of conflicting goals, such as productivity versus connectivity, which often jeopardize profitability and the ability to grow. What works well during bullish years may be ineffective during a subsequent downturn. In response, airlines and airport operators have learned to be flexible with costs in vital areas such as crew remuneration, fuel hedging, and aircraft financing. Revenue management has developed technologies, such as

[a] Networks are usually meant to provide proximity. Air traffic control, however, is an exception as it is designed to guarantee minimum distance.

sophisticated pricing mechanisms, that profitably respond to volatilities in demand while keeping capacity as tight as possible. It sometimes seems that a comparable level of conceptual and methodological sophistication in network strategy is compensated by rather expensive experimentation. This book aims to help in paving the way for more structured methodology in this field, initiating more relevant scientific work, and providing insight into the practical day-to-day work.

Airports no longer perceive themselves as mere providers of infrastructure. An increasing number of airport executives view themselves as network operators, with airlines serving as the suppliers of such networks. As a result, airports have significantly enhanced their capability to attract and acquire traffic, and have actively shaped their portfolio of destinations, frequencies, and passenger flows. To avoid being dependent on codeshares, airports have started to market interline connections. As airports begin to enter the arena, the management of connectivity—though small scale up till now—is no longer the sole privilege of airlines.

Hence, this book is written for a broad audience: experienced airline and airport network planners who may be looking for a new strategy or tactic; novices in network planning or market research departments who want to get up to speed; students of aviation management who must understand the wide spectrum of the subject; executives who are preparing for discussions with expert-level management; and regulators who recognize the complex economics of the aviation industry and the hidden effects of regulation.

By providing a conceptual framework for structuring this complex subject area, I hope to stimulate further focused and useful research. Although this book includes some mathematics, it remains within the scope of a high school curriculum to facilitate easy learning.

I am deeply grateful to the friends, colleagues, clients, family members, and others who helped to make this book possible. To name a few: Dr. Martina Bihn, Giorgio Callegari, Francesco Calvi, John Campbell, Mesut Cinar, Prof. Dr. Roland Conrady, Andreas Deistler, Harald Deprosse, Mariano Frey, Angelika, Matthias and Johannes Goedeking, Dr. Tobias Grosche, Cora Hartmann, Prof. Dr. Wolfgang König, Prof. Dr. Karsten Leibold, Kerstin Lösekamm, Maren Marczinowski, Annegret Reinhardt-Lehmann, Prof. Dr. Franz Rothlauf, Stefano Sala, Daniel Sallier, Tom Stalnaker, Kerstin and Pauline Strowa, Eva-Maria Sturm, Adam Seredynski, Linda Walsh, Andrew Watterson, and Wu Qi.

Unless specified otherwise, analyses are based on September 2009 data.

Frankfurt am Main, May 2010 Philipp Goedeking

Preface

This book is a much-needed, insightful synthesis of the latest knowledge and methodologies in network management. It covers a subject that is at the heart of strategy and operations at the Lufthansa Group, which today includes the networks of Swiss, Austrian, British Midland, Brussels Airlines, Air Dolomiti and Germanwings, serving more than 270 destinations and providing passengers nearly 1 million competitive and seamless transfer opportunities each month.

Deregulation in the United States and Europe replaced the traditional emphasis on nonstop routes with complex transfer origin-and-destination flows, paving the way to a new and augmented role for network management and revenue management. The concept of network management has led to an intrinsic integration of management of the portfolio of destinations and frequencies, pricing, crew and fleet management, and many other functional disciplines. Lufthansa has gone a step further, to the management of multiple networks. We devote a lot of attention to this issue, in order to keep organizational complexity at bay and unleash entrepreneurial energy.

The cyclical economic roller coaster in aviation has put many airlines in severe jeopardy in recent years. Even though most major airlines have learned to respond with early capacity cuts, each crisis creates the need to structurally realign the networks and resources of many airlines. Lufthansa itself went through this painful experience in the early 1990s, only to emerge as a much stronger airline. Swiss was able to restructure an over-sized network and fleet into a profitable airline, which is now an important part of the Lufthansa Group.

With our experience in managing and restructuring advanced networks, and in creating and expanding Star Alliance, Lufthansa is considered a global leader when it comes to modern network strategies and hub structures. Most of this expertise resides in the brains of our employees, as well as in organizational processes and procedures, and is scattered throughout thousands of documents. Yet surprisingly, there is no single monograph dedicated to modern network strategies and structures in our libraries or, to our knowledge, in any library around the world. This apparent vacuum is even more surprising when compared to the vast amount of literature on revenue management. The academic literature we find on network management focuses mostly on sophisticated computerized scheduling and resource allocation, but not on fundamental strategy.

This book fills the vacuum. Building upon his unique experience advising airlines, airports, and regulators all over the world in the area of network strategies and structures, Dr. Philipp Goedeking has written a thorough overview of the latest knowledge in network strategies and structures, covering all relevant aspects—from basic market research to the underlying math of sophisticated bank structures. The book is easy to read and maintains a thoroughly strategic perspective. Importantly, the author offers innovative perspectives on which direction network strategies and resulting structures might evolve—in everyday practice at airlines all around the world as well as in academic research.

As such, this book should be a must-read for everyone in the aviation industry.

Frankfurt am Main, May 2010

Dr. Christoph Franz
Deutsche Lufthansa AG
Deputy Chairman of the Executive Board
Chief Officer, Lufthansa Passenger Airlines

Contents

Abbreviations

IATA Airport and City Codes Used

AMS	Amsterdam-Schiphol Airport, Amsterdam, Netherlands
ATL	Atlanta International Airport, Atlanta, GA, US
AUH	Abu Dhabi International Airport, UAE
BOS	Boston Logan International Airport, Boston, MA, US
BRU	Brussels Airport, Belgium
BWI	Baltimore-Washington International Thurgood Marshall Airport, MD, US
CAN	Guangzhou Baiyun International Airport, China
CDG	Paris-Charles de Gaulle Airport, France
CVG	Cincinnati/Northern Kentucky International Airport, OH, US
DCA	Washington, DC-Ronald Reagan National Airport, VA, US
DEN	Denver International Airport, CO, US
DFW	Dallas-Fort Worth International Airport, TX, US
DOH	Doha International Airport, Qatar
DTW	Detroit Metropolitan Wayne County Airport, MI, US
DXB	Dubai International Airport, UAE
EWR	New York-Newark Liberty International Airport, NJ, US
FLR	Florence Amerigo Vespucci Airport, Italy
FRA	Frankfurt-am Main Airport, Germany
GRU	São Paulo-Guarulhos International Airport, Brazil
HAM	Hamburg Airport, Germany
HKG	Hong Kong International Airport, Hong Kong
HRG	Hurghada International Airport, Egypt
IAD	Washington, DC-Dulles International Airport, DC, US
IAH	Houston-George Bush Intercontinental Airport, TX, US
IST	Istanbul Atatürk Airport, TR
JFK	New York-John F. Kennedy International Airport, NY, US
JNB	Johannesburg-Johannesburg International Airport, South Africa
LAS	Las Vegas-McCarran International Airport, NV, US

LGA New York-LaGuardia Airport, NY, US
LGW London-Gatwick Airport, GB
LHR London-Heathrow Airport, GB
LON London, GB
MAN Manchester Airport, GB
MDW Chicago Midway International Airport, IL, US
MEM Memphis International Airport, TN, US
MSP Minneapolis-St. Paul International Airport, MN, US
MUC Munich-Franz Josef Strauss Airport, Germany
MXP Milan Malpensa Airport, Italy
NYC New York, US
ORD Chicago O'Hare International Airport, IL, US
PEK Beijing Capital International Airport, China
PHX Phoenix-Sky Harbor International Airport, AZ, US
PIT Pittsburgh International Airport, PA, US
PRG Prague Airport, Czech Republic
PVG Shanghai Pudong International Airport, China
RMF Marsa Alam International Airport, Egypt
SEA Seattle-Tacoma International Airport, WA, US
SHA Shanghai Hongqiao Airport, China
SIN Singapore Changi Airport, Singapore
STL Lambert-St. Louis International Airport, MO, US
TSN Tianjin Airport, China
VIE Vienna International Airport, Austria
ZRH Zurich-Kloten Airport, Switzerland

IATA Airline Codes Used

AA American Airlines, United States
AE Mandarin Airlines, Taiwan
AF Air France, France
CA Air China, China
CG Airlines of Papua New Guinea, Papua New Guinea
CI China Airlines, Taiwan
CO Continental Airlines, US
CX Cathay Pacific Airways, Hong Kong
CZ China Southern Airlines, China
DL Delta Air Lines, US
EK Emirates, UAE
KA Dragonair, Hong Kong
KL KLM, Netherlands
LH Deutsche Lufthansa, Germany
LUV Southwest Airlines, US
LX Swiss International Air Lines, Switzerland
LY El Al Israel Airlines, Israel

NW Northwest Airlines, US
OS Austrian Airlines, Austria
SN Brussels Airlines, Belgium
SQ Singapore Airlines, Singapore
TG Thai Airways, Thailand
UA United Airlines, US
US US Airways, US

IATA Country Codes Used

GB United Kingdom
TR Turkey
US United States

Other Abbreviations and Symbols

A Edges or arcs
A321 Airbus A321
A380 Airbus A380
ACI Airport Council International
APM Airline profitability model
ASK Available seat kilometers
ATI Antitrust immunity
b Number of banks
B737 Boeing 737-700
BDI Duration of inbound bank
BDO Duration of outbound bank
BDT Total duration of bank (inbound and outbound banks combined)
Bh Block hours
BOF Level of overlap between inbound and outbound banks
BSP Bank Settlement Plan
CAA UK Civil Air Transport Authority
CAAC Civil Aviation Administration of China
CB Connection builder
CDP Credit default probability
CDS Credit default swaps
CO_2 Carbon dioxide
CRS Computer reservation system
d Number of destinations
DB1B US Airline Origin and Destination Survey
DD Domestic-to-domestic connection
DFS Deutsche Flugsicherung GmbH
DI Domestic-to-international connection
dir Number of directional segments
DR Number of directional segments

DST	Generic three-letter code for destination
F	Frequencies
facc	Number of accessible directional outbound segments
fav	Number of available directional outbound segments
FiFo	First-in-First-out
GAO	US Accountability Office
GDS	Global distribution system
IATA	International Air Transport Association
ICAO	International Civil Aviation Organization
ID	International-to-domestic connection
II	International-to-international connection
IMF	International Monetary Fund
inb	Number of inbound movements
KPI	Key performance indicator
lag	Lag (in minutes) between corresponding inbound and out bound half-waves
LCC	Low-cost carrier
LiFo	Last-in-First-out
m	Economic "mass" or power
MaxCT	Maximum connecting time
MCT	Minimum connecting time
MIDT	Market information data tapes
min	Minutes
mio	Millions
MRO	Maintenance and repair organization
N	Network
n	Number of airports
NO_x	Nitrogen oxide
NPV	Net present value
O	Origin
O&D	Origin and destination
OAG	Official Airline Guide Corp
Ops	Operations
ORG	Generic three-letter code for origin
outb	Number of outbound movements
p	Probability of a hit
p.a.	Per annum
P&L	Profit and loss statement
P2P	Point to point
POS	Point of sale
PSO	Public service obligation
Q	Number of directional routes
QSI	Quality of Service Index
r	Directional reduction factor
R	Gravity

s	Number of directional outbound segments
shift	Time shift of outbound bank against corresponding inbound bank
SL	Stage length
SM	Short/medium haul
STA	Scheduled time of arrival
STD	Scheduled time of departure
T	Opening hours (in minutes) of an airport
t	Time
T100	US segment traffic data database
TAT	Turnaround time of aircraft
te	Excursion time
util	Utility
V	Vertices or nodes

If directions such as North, East, South, or West are abbreviated by capital letters N, E, S, or W, respectively, the text will explicitly point to this meaning.

Chapter 1
Market Research: Overcoming Incomplete, Inconsistent, or Outdated Data

Abstract Strategy is the art of asking, "Why?" To determine why a particular network of destinations is being served, two questions are key: (1) What are the level, structure, and evolution of demand on selected parts of the network?, and (2) What are the strengths, weaknesses, opportunities, and threats of the various market participants on those markets? Reliable answers to both questions require extensive market research. While many institutions provide data on demand and supply for local markets, the more networked and interdependent effects on, for instance, city pairs may be difficult to research, inconsistent, missing for selected markets, outdated, or prohibitively expensive to regularly acquire. Market research in aviation is becoming increasingly complex. The aviation industry is providing large numbers of publicly or commercially available figures on many market aspects: volume and evolution of demand or supply, level of competition, and operational or financial results of airlines and other players in the aviation value chain. However, the various sources of data are frequently deficient or suffer from significant missing values or inconsistent definitions. In contrast to the plethora of available data, much vital information must be researched manually. This, in turn, depends on individual access to corporate decision makers or governmental bodies. In this chapter, a framework of industry terminology is offered along with a discussion of the most important sources of data in this field. Various perspectives on aviation markets also are reviewed.

1.1 A Quick Review of Fundamental Theory

Much of the theory on networks is based on mathematical graph theory. Essentially, graph theory offers a conceptual framework that is intuitively understandable. Simple and straightforward language is used to describe various aspects of aviation networks. Graph theory also helps to visualize important aspects of networks. A basic understanding of graph theory is critical in understanding the structure and dynamics of airline networks.

P. Goedeking, *Networks in Aviation*, DOI: 10.1007/978-3-642-13764-8_1,

First, what is a network? Formally, networks are defined as

$$\text{Network} = \text{objects} + \text{connections (Casti 1995).}$$

A network N can easily be understood as a set **V** of vertices or nodes, and a set **A** of edges or arcs. Since edges can only be constructed between nodes, the maximum number of edges in any given network is the Cartesian product of $V * V$. Hence, any real set A of edges in a network is at least a subset of $V * V$.

$$\begin{aligned}\text{Network N} : &\text{ set } \mathbf{V} \text{ ofvertices or nodes}\\ &\text{ set } \mathbf{A} \text{ of edges or arcs}\end{aligned}$$

where $A \in V * V$.

While edges always represent direct links between two nodes, paths are defined as a sequence of edges traveling through the network with no edge being used more than once. Different from a path, a circle is a sequence of edges in which the first and the last nodes are the same.

Let us assume a simplistic network of nine nodes, A to I.

Figure 1 gives an example of a set of possible edges in that network, clearly a subset of all theoretically possible connections. Each edge connects two nodes. In airline networks, the nodes represent airports and the edges indicate flights. In this example (1a), all edges are nondirectional. In airline networks, this is usually not the case. Passengers want to fly in one direction—the return flight would be a directional edge (or path) pointing into the opposite direction. To be more realistic, we give each individual edge a direction, as shown in Fig. 1b. Passengers who want to fly from node "G" to node "A" do not find an edge connecting these two nodes, or airports. Instead, these passengers have to fly a series of subsequent edges, or a path. Figure 1c shows the directional path from G to A, composed of a subsequent series of edges. In airline language, the respective passengers would have to connect. The definition given above of a path (no edge being used more than once) is fully satisfied in this example. In Fig. 1d, two kinds of circles are given. The circle starting in A and then looping through D, G, H, E, B, and back to A is a multi-segment circle. The circle on the right-hand side, looping ping-pong between nodes C and F, is the smallest possible circle in a network. Figure 7 illustrates how extensive circles can be found in airline networks and in ping-pong patterns.

The term "path" is used in a broader sense to include both the strict definition of paths as well as circles.

1.2 What is an Aviation Market?

In aviation, the term "market" has two meanings. Market may refer to a region, vicinity of an airport, city, or catchment area as the origin or destination of traffic. The same regional market may be a "source" market for outbound demand, or it may be a "destination" market for inbound traffic. Alternatively, the term market

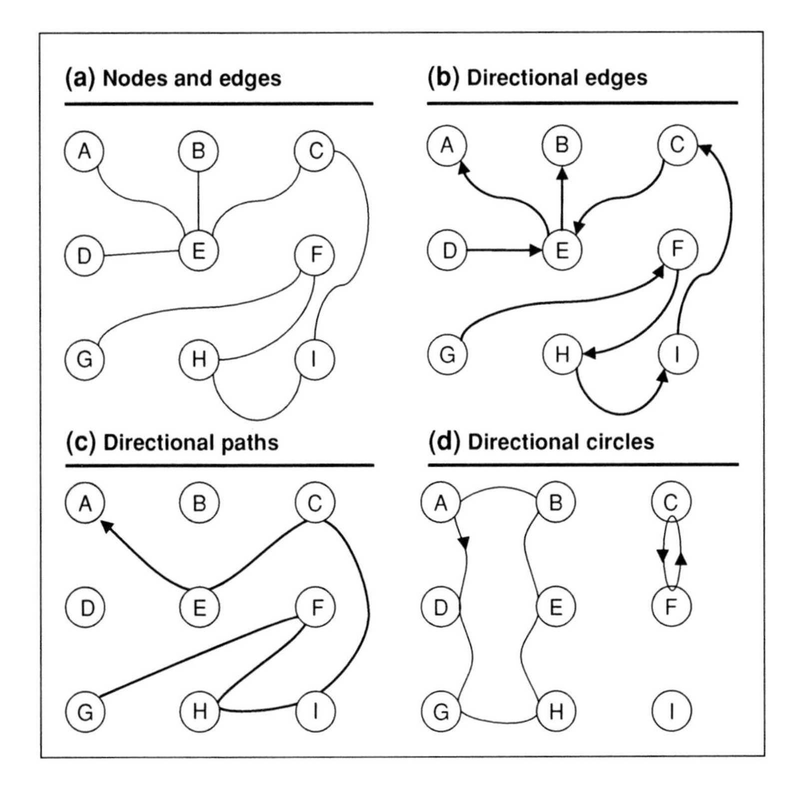

Fig. 1 Network prototypes. A simplistic network with (**a**) nodes and edges, (**b**) directional edges, (**c**) directional paths, and (**d**) directional circles

may refer to the relationship between two regions or cities or to the edges. Both aspects require different data and methodologies to research demand, supply, competition, and the respective regulatory framework.

To give an example, there is a strong source market in the New York metropolitan area (NYC). The JFK market differs from the EWR market, and so does the market between Berlin/Germany and New York. Market research in aviation must comprehend the local (node) as well as the city pair (edge) dimension of the term market. Market can also be applied to a broader set of nodes and/or city pairs. For instance, the Mexican aviation market summarizes all traffic to and from Mexico City, Cancun, and other cities; while the Italian market comprises all traffic to or from Rome, Milan, Palermo, and other Italian cities. The Mexican and Italian markets can be conceptualized as meta-nodes that conceal the traffic within the node, leaving only some edges (international traffic) connecting the meta-node with the outside international network.[1]

[1] The concept of meta-nodes can be compared with the fundamentals of software engineering. McMenamin and Palmer (1988) describe the idea of meta-nodes to frame a group of processes into a meta-process—this technique is also known as "bubble charting."

A simple example (see Fig. 3) should help clarify some of this terminology. Remember that traffic is always directional and that such directionality can be represented as directional arches between nodes. Hence, the first differentiation we can make is to distinguish whether a particular region is the origin or the destination of a passenger's journey. We will briefly discuss both instances.

1.3 Outbound Markets

The ticket for the itinerary FRA–LHR–FRA (see Fig. 3) is sold for departure in FRA. Often, "FRA" is considered the "point of sale," or POS, of such a ticket, but this is not necessarily correct. Consider the following hierarchy:

A catchment area is the broadest concept of the regional vicinity of a given airport. Airports and airlines sometimes define their catchment area by the population that can reach the airport within a 1 or 2-h drive by car or train, or within 100–200 km of distance. The purpose of the catchment area concept is to understand the population that a given airport can draw upon as potential passengers.

The effective source market of an airport refers to the number of embarking passengers per month or year for whom the given airport is the first point of their journey or of their return journey. The catchment area points to the potential market dimension, while the effective source market refers to the actual level of exploiting the potential.

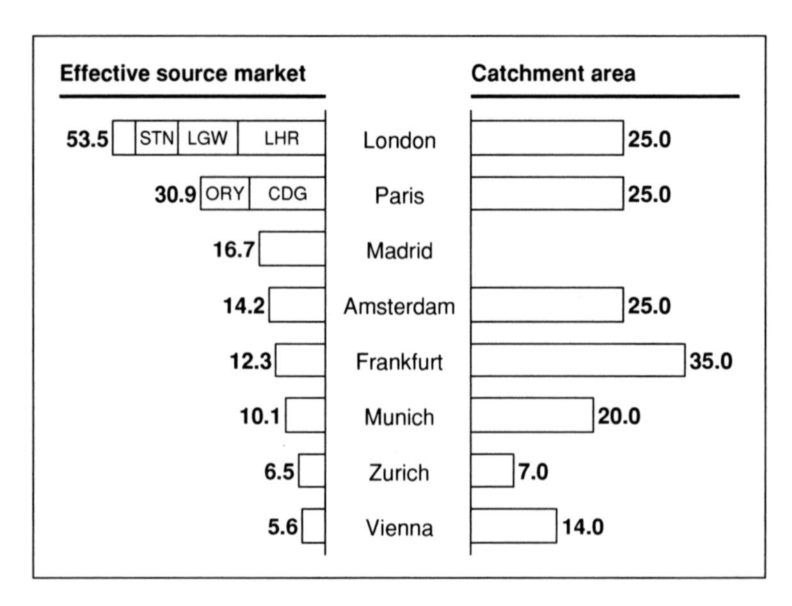

Fig. 2 Effective source market size and catchment area for selected European hubs [mio passengers p.a.]. The size of the MAD catchment area is not known (Goedeking et al. 2009)

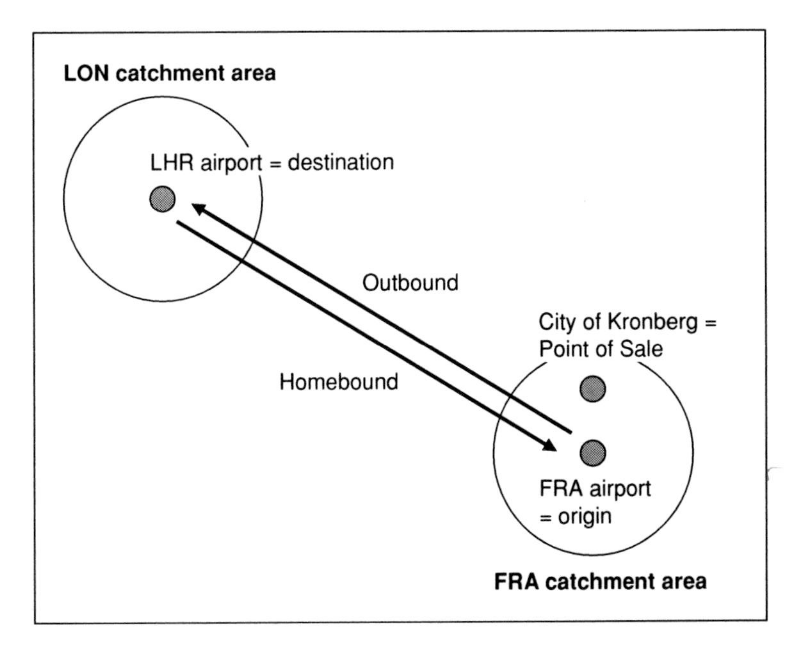

Fig. 3 Terminology of source and destination traffic. 1. A business class ticket FRA–LHR–FRA is sold in Kronberg, making Kronberg the point of sale (or POS) and source market for this itinerary. 2. The itinerary starts in FRA, making FRA the origin of this itinerary. Note that POS and origin are not necessarily identical. 3. LHR is the destination of the itinerary at hand. From the point of view of this itinerary, London is an inbound market

Figure 2 illustrates the significant differences between the size of a catchment area and the dimension of the corresponding effective source market for some selected European hub markets.

The Point of Sale, or POS, of a ticket refers to the most detailed level of specificity of a source market. The POS may refer to a specific city, community, or travel agency selling this ticket.

To give an example (see Fig. 3): a business class ticket FRA–LHR–FRA is sold by the agency Hessen-Travel GmbH in Kronberg, a small city near Frankfurt am Main, Germany. Kronberg and the Hessen-Travel agency represent the POS of this ticket; FRA airport is the origin of the ticket; and the catchment area of FRA extends as far as Kronberg. Note that in London's passenger statistics, this ticket would be counted as inbound traffic.

1.4 Inbound Markets

Some airports, such as Hurghada (HRG) in Egypt, serve almost exclusively as tourist destinations. Those destination airports are referred to as "inbound" markets. For inbound markets, the proportions are significantly more complex and difficult to assess than for outbound (or source) markets. Passengers flying to HRG might continue by car or other modes of transportation to a final destination outside HRG or Egypt. Such information cannot be deduced from any data available in inventories. To understand the true nature of directional O&D demand, one must understand the final destination of a journey (not merely the final airport of a multimodal itinerary), as it is important to know the genuine POS of a ticket. Even though some advanced operations-research science is examining the most important multimodal itineraries in a particular region around a given airport (Mandel et al. 1997), the only reliable way to understand such particulars of inbound markets is by interviewing passengers. Several years ago, such customized market research led to the erection of a new airport in the south of Egypt, close to the Sudanese border (Marsa Alam, RMF), in order to drive tourist development.

The differentiation between inbound and outbound markets is crucial because certain airline strategies can be applied more effectively to either outbound or inbound markets. Hubs are airports with a high volume and proportion of transfer traffic. Because major hubs rarely succeed in strong inbound markets, the majority of hubs are based in pronounced outbound markets around the globe. Strong outbound markets are usually driven by a high proportion of time-sensitive and high-yield demand, providing sufficient backbone traffic to permit the development of lower-yield transfer traffic. Destinations like Hurghada in Egypt (HRG) or other tourist destinations around the globe are mostly inbound markets. Very few passengers buy their tickets in HRG. However, a rapidly growing number of passengers want to fly to HRG and therefore buy their return tickets to HRG in points of sale around the globe, predominantly in Europe and Russia. Attempts to upgrade even sizeable inbound airports into transfer hubs are likely to fail.

Special cases are hubs like Dubai (DXB), Abu Dhabi (AUH), or Doha (DOH) in the Gulf region. These hubs are neither strong outbound nor strong inbound markets. Instead, the business model of these hubs is built upon high volumes of transfer traffic. What sets them apart from hubs in Europe is the high share of longhaul-to-longhaul connections. Of all transfer connection opportunities (or "hits," see Chap. 2) in DXB, more than 26% are longhaul-to-longhaul. The comparative figures for mature hubs like CDG and LHR are 1 and 6%, respectively.

1.5 O&Ds, Routes, and Flights

Consider a passenger wishing to fly from MAN to IST (Manchester/GB to Istanbul/TR). The passenger's primary demand is MAN–IST, and the choice of a specific path or a specific connection is secondary and constrained by available supply. The strict passenger perspective—viewing the genuine origin and destination as a market regardless of actual production issues—is called an Origin-and-Destination market, abbreviated as O&D. Airports frequently use the term O&D with a different meaning. For airports, an O&D is usually all traffic with a given airport as the genuine starting point of a nonstop or multi-segment journey, or the final destination of a nonstop or multi-segment trip originating elsewhere in the world. Hence, what is an O&D from one airport's point of view is a connect market from another airport's perspective.

To avoid this ambiguity, we will stick to the passenger point of view, defining an O&D as the market between the first and final point of a trip (see Fig. 4). Since some cities have several airports due to infrastructure constraints, an O&D is between cities rather than airports. In some large metropolitan areas, however, airports within the same metropolitan area may serve different regional markets, such as JFK and EWR in the NYC metropolitan area. In most cases, however, the split of traffic between multiple airports in a given metropolitan area is regulated, not market driven, as in Shanghai Pudong (PVG) and Shanghai Hongqiao (SHA).

In the schematic example given in Fig. 4, there are eight possible paths or itineraries between MAN and IST, seven of which must connect (via LHR, AMS, FRA, PRG, CDG, ZRH, or MUC) and one nonstop. An itinerary is defined as a nonstop or connecting path through a network to travel from the origin to the destination. Note that an itinerary connects an airport pair not a city pair. The O&D NYC-LON can be served by at least three different nonstop itineraries: EWR-LHR, JFK-LHR, and JFK-LGW.

A route is a nonstop itinerary between two airports. Flights typically carry a flight number. Note that quite often a sequence of flights carries a common (virtual) flight number (Holloway 2003). Although most flights carry a unique flight

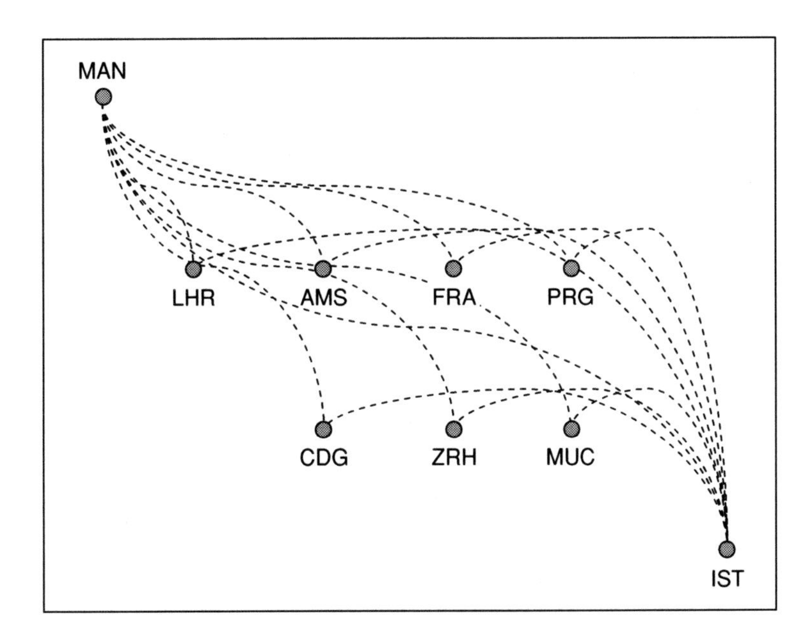

Fig. 4 Routes, itineraries, and O&Ds. The market MAN–IST is defined as an O&D. The O&D market MAN–IST can be served through eight various itineraries; one is nonstop and the others are connect. The itinerary MAN–LHR–IST is served by two subsequent routes

number, not every flight number points to a unique flight. The terms "leg" and "segment" refer to the operational nonstop constituents of a route or flight. The itinerary MAN–AMS–IST in Fig. 4 is composed of two connected routes: MAN–AMS and AMS–IST.

The various terminologies relating to transfer connections are inconsistent. A qualified connection (see Sect. 2.3.1 in Chap. 2) is a commercially reasonable, competitive, operationally feasible, and sufficiently convenient transfer opportunity between arriving and departing flights. We will use the term "hit" for such qualified connections. Hits, together with routes, represent the constituents of itineraries.

Airports use additional terminology to differentiate between various kinds of connections; these terms are summarized in Fig. 5. Airports or hub managers at airlines (see Sect. 6.4 in Chap. 6) often need to view transfer flows from the perspective of a particular airport rather than from the demand-driven passenger point of view. A nonstop route between two airports (hubs A and B, see Fig. 5) is often referred to as point-to-point, or P2P. From hub A's perspective, a "behind" flow is arriving from any airport(s), connecting in A to connect to B. The "beyond" originates in A, flies to B, and connects in B to any beyond destinations. Finally, a "bridge" flow refers to flows originating wherever, connecting in A to B, and connecting in B to wherever.

This airport-specific perspective reflects the fact that airports are not primarily interested in understanding the true O&D flows, but in understanding the actual flows through their respective airport. The question they ask is: Where do the passengers come from who connect at my airport, and where are they going to next? Airports need to understand how each transfer flows through their airport at the level of individual passengers as a prerequisite for charging and accounting for fees to the airlines.

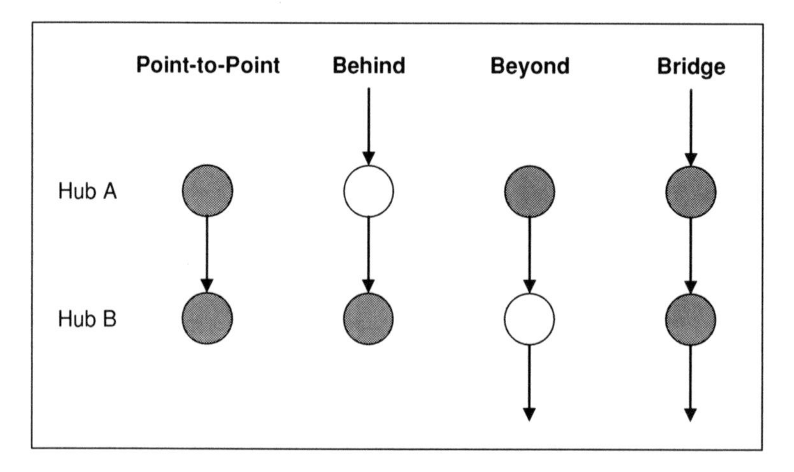

Fig. 5 Point-to-points, behinds, beyonds, and bridges

1.6 Accessing Essential Market Research Data

Obtaining the required data to correctly assess the inbound/outbound proportions for airports worldwide is not a simple process. Transaction data from the global distribution systems (GDS) or booking data from International Air Transport Association (IATA)-certified sales agents (BSP data, Bank Settlement Program) may correctly reflect most outbound markets, as outbound markets usually build upon large cities and strong business traffic. However, strong inbound markets like Antalya in Turkey are mainly served by leisure airlines that rarely sell through GDS or IATA agents. It is difficult, if not impossible, to obtain reliable passenger statistics for strong inbound markets and tourist markets, in particular. As a rule of thumb, the capacity share of the largest airline serving a particular airport can be used as an indicator of inbound/outbound proportions. Pure tourist or other inbound markets are characterized by a fragmented pattern of airlines serving this destination, while strong outbound markets are usually dominated by a "home carrier."

1.7 Researching Local Aviation Markets

To get an overview of any local market, the following dimensions must be addressed:

- Volume and structure of demand
- Competition—competing nonstop and/or transfer itineraries
- Infrastructure capacity and throughput
- Regulatory framework such as traffic rights.

In Local Markets (see Appendix: "Market research checklist"), a more detailed checklist is provided. Market research for local markets, in particular, depends on local data and word-of-mouth information. These globally available aviation data sources will prove helpful:

- ACI (Airport Council International)
- GDS transaction data (also referred to as Market Information Data Tapes, or MIDT)
- BSP data (IATA)
- National or governmental statistics, such as the European Statistical Office (Eurostat), T100, DB1B, Civil Aviation Administration of China (CAAC), and the GB Civil Aviation Authority (CAA)
- Airline and airport websites
- Airline and airport reports
- Electronic flight schedules.

Experienced planners also rely on qualitative information. When companies open new factories or shut down existing ones, or when they acquire or sell other firms, their corporate plans may contain useful hints on future developments of specific routes and markets.

Practical tips on the volatility of demand in a particular country can be derived from the interest payable to long-term private sector or sovereign credits, called CDS spreads (credit default swaps), particularly if CDS spreads are converted into default probabilities (CDP, credit default probability). High-default probabilities point to a risk of high-demand volatility, whereas low-default probabilities suggest a stable economy and sustainable demand. Other sources of market risk may be obtained from the IMF (International Monetary Fund) or the World Economic Forum (Schwab 2009).

1.8 Researching City-Pair Markets

City pairs may be connected either nonstop (edges) or by transfer connections (paths covering more than one edge). It is important to understand that the issue of nonstop versus connection is purely a matter of supply, not demand. If the demand between two cities is served nonstop, such markets are called nonstop or P2P (see Sect. 1.5). Note that the term "local" refers to a POS or particular airport (nodes), but not to city or airport pairs (edges). City pairs requiring transfer connections are referred to as transfer or connecting traffic. The vast majority of city pairs (also referred to as Origin-and-Destination, or O&D markets) are supplied as transfer connections. Researching the demand characteristics of multi-leg city pairs is much more complex than understanding local point-to-point or nonstop markets.

Another important aspect of researching city-pair markets is recognizing the directional nature of city pairs: the market A-B may be quite different from the market B-A.

Researching city pair or O&D markets is more complex than doing so for local markets. Data availability is more constrained, and accessing data is more costly and more difficult due to language barriers. The following key issues are critical in understanding city-pair markets:

- Volume and structure of demand, including breakdown by service and/or fare class
- Competition, such as competing transfer itineraries
- Macroeconomics of origin and destination
- Regulatory framework, such as traffic rights.

City-pair market data are much more difficult to obtain and usually more expensive to acquire than airport or city-specific data. Here are some important sources for city-pair market data:

- GDS transaction data—also referred to as MIDT (see above)
- BSP data (IATA)

- National or governmental statistics
- International Civil Aviation Organization (ICAO)
- Airline and airport reports
- Electronic flight schedules
- Commercial vendors of O&D data.

The advent of global distribution systems (GDSs) in the late 1980s and early 1990s offered an ideal opportunity to obtain relevant reflections of demand volumes and structures on multiple-stop itineraries. Managing 90% or more of all booking transactions in Europe, North America, and most of the Asian markets, the GDSs began to market these transaction data as "Market Information Data Tapes," or MIDT.

Airlines subscribed to the major GDSs to acquire a copy of their booking transactions. In return, they received a detailed picture of which city pairs had been booked during a particular time period, which flights were chosen for nonstop or connect, which agency in which city booked the flight, and how long in advance of the day of operations the respective flight was booked. Numerous software companies developed and marketed sophisticated tools to provide detailed and informative reports of BSP or MIDT data. MIDT data were invaluable for sound network planning as well as for sales planning and control, such as agent bonus programs.

Yet, these "good old days" appear to be over (Sala 2009). With the sharp increase of direct sales by airlines, in particular low-cost carriers (LCCs) through the Internet and call centers, these bookings no longer appear in the GDSs. As a result, the MIDT data are no longer as valid and complete. New legislation[2] and court decisions,[3] at least in Europe, further constrain the use of key elements of transaction data, such as the "agent code" that identifies the issuing agent of a ticket.

To overcome the apparent vacuum of market intelligence on city-pair demand, airlines, research institutions, and market intelligence firms have developed a variety of methodologies to cope with this challenge. The following is a brief review of the four most promising approaches:

1.8.1 Traffic Projections Based on Time Series

An O&D that provides large volumes of traffic for an extended period of time is unlikely to implode suddenly or to inflate abruptly in demand volume. Hence, many market researchers apply time series based forecasting techniques to project current figures from historic data. Time series traffic projections assume that, to

[2] Directive 96/9/EC of the European Parliament and of the Council of March 11, 1996, on the legal protection of databases.

[3] Amadeus versus IATA International Chamber of Commerce arbitration, May 2009.

some extent, the future is a function of the past. Because this assumption too often proves to be wrong, time series based techniques have lost acceptance in the aviation industry.

1.8.2 Upscaling

Many researchers use BSP and/or MIDT data to scale up more realistic figures of market size. One airline, for instance, simply multiplies MIDT figures by 1.4 to come up with a more realistic market size figure. Typically, airlines or market research companies use additional data sources to estimate, ranging from their own inventory data to telephone traffic data. However, the multipliers applied to upscaling are estimates themselves, and can hardly be calibrated to provide sufficient significance. An already deficient MIDT or BSP figure, multiplied by a fuzzy multiplier, is unlikely to offer precision as errors accumulate.

1.8.3 Gravitation Models

Some airlines and researchers use econometric models, such as gravitation models, to assess the demand volume between cities. These models are based on Isaac Newton's discovery that the gravity R of two masses m_1 and m_2 and the distance r in between them follows the rule

$$R = \frac{G(m_1 * m_2)}{r^2} \tag{1}$$

where $G = 6.67 \times 10^{-11}\ \mathrm{Nm^2/kg^2}$.

The hypothesis of gravitational models in transportation is that the economic power of a city or catchment area can be conceptualized as its "mass" (Grosche and Rothlauf 2007). Following this hypothesis, passenger demand in between two cities or catchment areas should be in proportion to the "economic mass" m_1 of city 1 and the "economic mass" m_2 of city 2; and it should be in reverse proportion to the distance in between them. An excellent indicator of the economic power ("mass") of a city is the seat capacity offered at its airport(s). Many other factors contributing to the "gravity" of economic masses—such as direct seats or frequencies between m_1 and m_2, common nationality, common markets or currencies, fixed or mobile telephone traffic, Internet traffic, common language, and common history—may be added to the formula.

Results from gravitation models should not be expected to provide reasonable outcomes at the level of individual O&Ds. They may, however, provide estimates that permit relative comparisons and rankings at an aggregated country-to-country level.

1.8.4 Reverse Engineering

A recently introduced technique re-engineers demand data from available passenger and traffic statistics. Starting with random assumptions on available demand for all relevant O&Ds worldwide, advanced optimization techniques are applied to find the single set of O&D demand estimates that can best explain all available passenger and traffic statistics.

1.8.5 White Spaces and Inconsistencies

At a first glance, the aviation industry appears highly transparent with a wide variety of data on demand and supply from virtually all markets worldwide. At a second glance, however, many significant data deficiencies become obvious.

1.8.5.1 White Spaces

While some markets are well covered by data, other market data availability is limited or data are difficult to access. In some regions around the globe, airlines do not upload their flight schedules to the vendors of electronic flight schedules or global GDSs. Hence, data on capacity deployed are not available for those markets. The booking transactions of some important GDSs are not commercially available, again creating severe white spaces in understanding the demand side. BSP data used for market intelligence do not cover some important markets, or are blocked for such application by court ruling.

1.8.5.2 Data Inconsistencies

When searching for the same data item in different sources of data, one easily finds a frustrating degree of inconsistency. For many US domestic routes, the two key data sources published by the US Department of Transportation, T100 and DB1B, provide significantly different data, which again do not match capacity data published in the electronic flight schedules. For example: On one selected US domestic route and period of time, DB1B states 260,000 passengers, T100 states 320,000, and flight schedules state a capacity of 470,000 seats. The German Statistical Office (Destatis) publishes a demand of 447 passengers for the nonstop O&D from FRA to a major transfer hub in Southeast Asia for a particular time period, while stating a volume of almost 24,000 passengers to fly transfer on the respective flight. For most airports around the globe, the published flight schedules are not balanced, so that schedules can claim more arrival than departure capacity, or vice versa. In some cases, the difference is fourfold. For leisure destinations, passenger figures vary significantly across various data sources. A typical leisure airport in Greece publishes a volume of 450,000 passengers, the European Statistical Office (Eurostat) states 170,000 passengers annually, and the electronic flight schedules announce available capacity of only 110,000 passengers.

1.8.5.3 Inconsistent Definitions

Should a transfer passenger count once at the point of transfer, or count twice when disembarking and embarking again? The international standard is to count transfer passengers twice. However, not all airports that publish transfer statistics apply this definition or specify which definition they apply. In some cases, airports change the definition without letting the public know. Another example of inconsistent definition refers to the different meaning of "Origin-and-Destination" (O&D) that is used among airline and airport personnel (see Sect. 1.5).

1.8.5.4 Lack of Credibility

Some airports publish data that are obviously incorrect or fabricated. For example, one major European airport for years has been publishing the same absolute figure of transferring passengers.

1.8.5.5 Inconsistent Data Periods

Some data are published daily (electronic flight schedules and MIDT transaction data); monthly (most airport reports); quarterly (reports of at least stock-listed airlines and airports); bi-annually (industrial associations); or annually (many governmental statistics). Some of these data are published immediately (electronic flight schedules) or up to a year later (some governmental statistics). To compare data stemming from sources covering various time periods, conversions sometimes must be made based on questionable assumptions. Traffic data for the first week of December for many markets worldwide, for example, cannot be extrapolated to a full month by dividing by 7 and then multiplying by 31, because December typically shows strong seasonal effects that are not yet at full strength during the first week of this month. As a minimum requirement, extrapolations must be based on careful analyses of capacity development during the specific time period.

Chapter 2
Network Structures Follow Network Strategies

Abstract In this chapter, we summarize and compare the structural, economic, and strategic rationales of hub-and-spoke and point-to-point network architectures; introduce the operational basics of aviation network structures; and outline the key tools needed to master complex planning and the controlling of aviation networks.

Historically, aviation networks were decentralized, regional structures. Airlines were government owned—hence the terms "national" or "flag" carrier—and served the public or political desire to provide air transport infrastructure, all based on tight regulatory rules and restrictions. After the US deregulation in 1978, highly centralized hub-and-spoke networks quickly emerged as the seemingly perfect answer to serve big markets at low costs. Rebuffing this strategy, LCCs mushroomed to attack the hub-and-spoke networks by offering direct services where the hub-and-spoke carriers only provided transfer services. Today, hub-and-spoke network architectures are playing the strengths of their network structures against LCCs, while LCCs have widely filled the vacuum of non-hubbed routes left by the hub-and-spoke carriers. Some of the largest hub-and-spoke carriers have adopted LCC hub structures, while maintaining the hub-and-spoke structure of their overall networks. In turn, LCCs are serving more transfer traffic and incorporating advanced revenue management systems to exploit the revenue potential from transfer traffic. LCC role model Southwest (WN) (Maxon 2010) is developing bank structures in Phoenix (PHX), Baltimore (BWI), Las Vegas (LAS), St. Louis (STL), Denver (DEN), and Chicago (MDW) to improve connectivity and transfer traffic. Both network structures and related business models are increasingly converging (Handelsblatt 2009). After many failed attempts, the first successful entries of LCCs on long-haul markets can be observed, while many hub-and-spoke carriers have launched LCC subsidiaries.

P. Goedeking, *Networks in Aviation*, DOI: 10.1007/978-3-642-13764-8_2,

2.1 Complying with Basic Operational Rules

Operational timing issues affect the feasible or desirable network structure of all airline networks, regardless of hub-and-spoke, LCC, or other strategic objectives behind a particular network. Therefore, we will review those operational issues—mostly rotational requirements—common to all structural network variants before studying the structural drivers of connectivity and productivity. While the differences and similarities of connectivity and production-driven network structures represent a continuum rather than distinct categories, the drivers of this continuum—connectivity versus operational efficiency—are clearly distinct in terms of strategic objective and operational levers. The following sections focus on understanding these underlying drivers rather than the resulting characteristics.

2.2 Sequencing Flights into Rotations

Passengers want to fly from A to B at particular times. Unfortunately, an airline cannot accommodate all passenger-timing preferences at the same time due to financial considerations or operational constraints. While trying to adhere to passenger demand as closely as possible, aircraft follow their specific paths through the network, mainly according to operational criteria, as do the cockpit or cabin crew. To describe such paths of aircraft, the term "rotation" is used. A rotation is a path characterized by the same airport serving as the start and final point. It may be composed of a single circle, but also may include multiple circles and other forms of paths. While the rules to build rotations for aircraft are purely technical, cockpit and cabin crews negotiate their rules in lengthy labor disputes with airline management.

Production requirements do not allow network structures to be truly distributed, but real-world networks do have centers. The crew has a home base and wishes to return to it as regularly as possible. Aircraft and maintenance facilities also are assigned to home bases. As a result, the centripetal forces of such bases create some form of centers within an airline network.

For many, the sequence of individual flights (edges) flown by a specific aircraft, along with the arrival and departure time for each destination (node), is equivalent to "the schedule." Such "rotational plans" are typically visualized by means of Gantt-charts (see Fig. 8). Similar rotational Gantt-charts are generated for cockpit or cabin crews.

In Sect. 1.1 in Chap. 1, two theoretical types of subsequent edges were differentiated: circles (first and last nodes are the same) or paths (no more than one edge). We find paths as well as circles in airline networks. In production, circles clearly prevail, since the aircraft and crew must eventually return to their home bases.

The most common type of circle is a "ping-pong" flight. A particular aircraft flies a route in one particular direction first, and then returns to the originating point of departure. This scheme has the advantage of high operational stability:

- Aircraft frequently loop through their home base to facilitate minor maintenance work.
- Aircraft of the same type and stationed at the same base all regularly arrive or depart at short time intervals at their home base; that way, one aircraft can easily be replaced by another, making operational disruptions less likely (see Fig. 6).
- A spare aircraft at the base can be used to remedy an operational disruption.

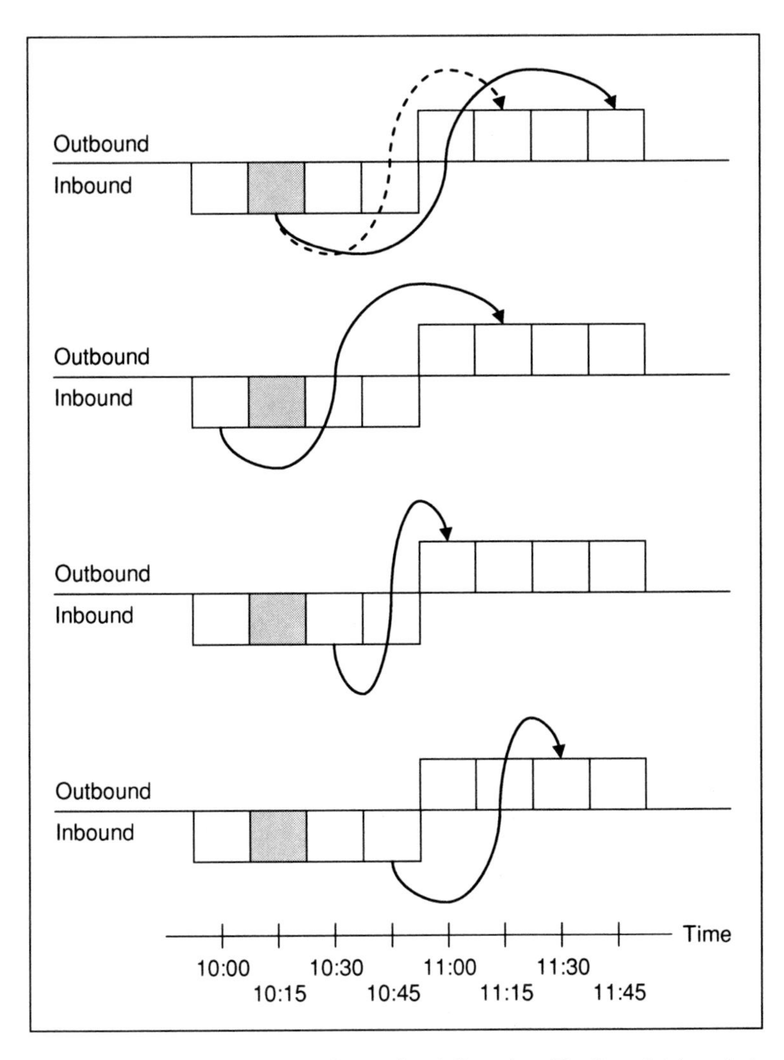

Fig. 6 Circular swapping in the event of operational disruption. The "*gray*" inbound aircraft is scheduled to depart as the second departure within the outbound wave (*dotted line*). Assuming a technical problem, this aircraft can take over only the last departure flight out of that wave (*solid line*). In this case, the sequence of rotational switches indicated in the *last three rows* can make sure all flights depart as scheduled [assuming a minimum ground time ("turnaround time," or TAT, see Sect. 2.2.1) of 30 min]

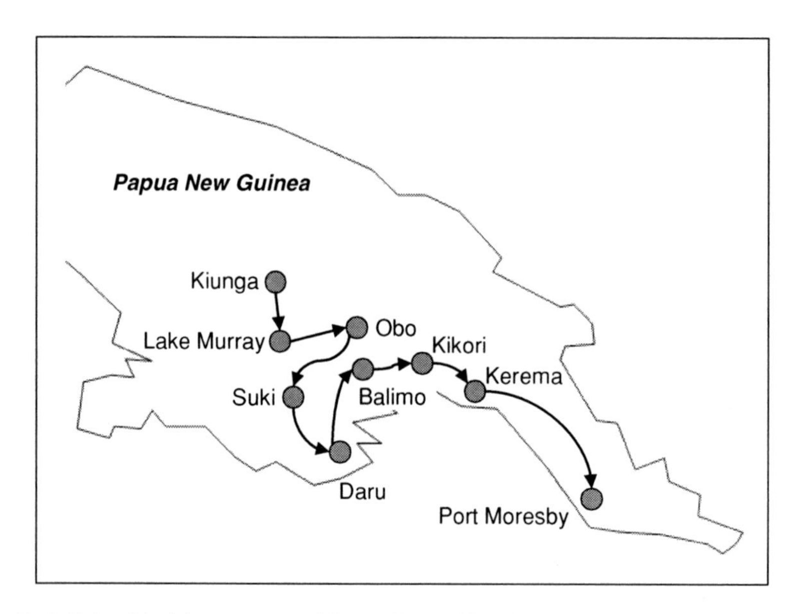

Fig. 7 A flight with eight segments. Airlines of Papua New Guinea flight number CD 359 covers a sequence of nine destinations under one single flight number

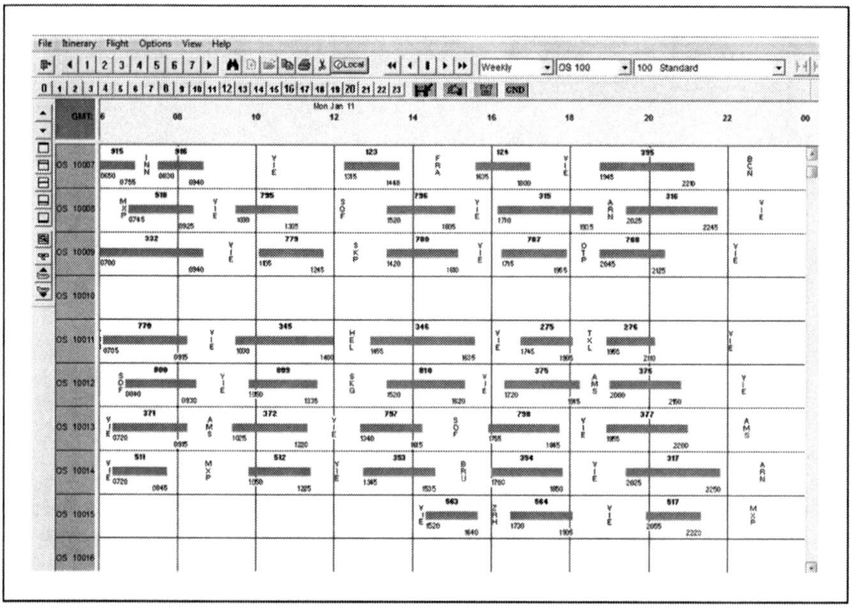

Fig. 8 A rotational map for an OS F100 aircraft (Plancor SA)

LCCs typically fly ping-pong due to their efficiency and robustness. At the other extreme, some airlines fly extended paths or rotations. A rather impressive example of such a "pearl collar" path of flights can be found on Papua New Guinea, where Airlines of Papua New Guinea (CG) serves, under one flight number (CG 359), a sequence of nine destinations (see Fig. 7), all on day seven (only).

2.2.1 Turnaround Time is Non-Productive Time

Aircraft need time after arrival before they can depart again. The time span from touching the gate ("on blocks") until pushing back from the gate again ("off blocks") is called the turnaround time, or TAT, of an aircraft. TATs are aircraft, airport, airline, and rotation specific. LCCs are famous for their highly efficient turnaround procedures, significantly increasing aircraft as well as terminal and apron asset utilization. TATs for a typical medium-haul aircraft such as an A320 or B737 lie in the order of 45 min, with smaller aircraft typically turning around faster, and wide-body aircraft requiring significantly more time. Some LCCs, such as Ryanair, turn a B737 around in as short a time as 20 min. Turns at an outstation are typically faster than turns at a base, where routine maintenance work is performed. TAT is different from "airport slack" (see Sect. 3.7 in Chap. 3), which is much broader and includes all time needed for taxiing, take-off/landing, and approach/climbing.

2.2.2 Building Sequences of Flights: FiFo and LiFo

Airlines typically apply two distinct methods to achieve efficient resource allocation: first-in-first-out (FiFo) and last-in-first-out (LiFo).

2.2.2.1 FiFo

Let us assume a string of inbound and outbound flights at a hypothetical airport (see Fig. 9). The task is to link ("rotate[4]") the B737 flights in an optimal way. FiFo simply takes the first inbound flight in the morning and links it with the first available B737 outbound. Then the same procedure is repeated with the next inbound flight until all the B737 flights are exhaustively linked. Of course, once an

[4] This kind of sequencing does not necessarily lead to a complete rotation in the sense of having a starting and final airport in common. In colloquial airline terminology, however, almost any sequencing of flights is referred to as "rotating."

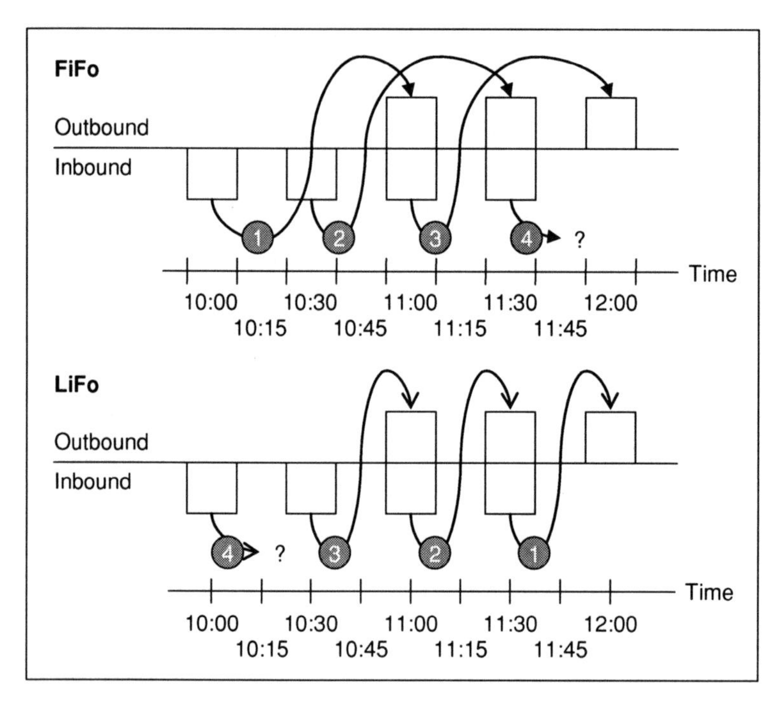

Fig. 9 FiFo and LiFo sequencing of flights

outbound flight is linked with an inbound flight the first time, it is no longer available for other links.

2.2.2.2 LiFo

Using the same example as in the FiFo case, LiFo starts with the last inbound flight (out of the B737 fleet) in the evening, linking it with the next available B737 outbound flight. The process then continues with the second last inbound, which is again connected with its respective next available (and not yet linked) outbound flight, and so on.

FiFo generates efficient, evenly packed rotational patterns; LiFo packs tighter, but sometimes opens up space. As a result, LiFo-based rotations may offer more, though limited, room to maneuver when flexibility is needed.

2.3 Hub-and-Spoke: The Answer to Deregulation?

Network airlines try to optimize connectivity, or the ability to arrive on one flight (edge) and connect to an outbound flight, thus creating a path. To do so, highly

Fig. 10 In a point-to-point network, the connection of 5 nodes requires 20 flights

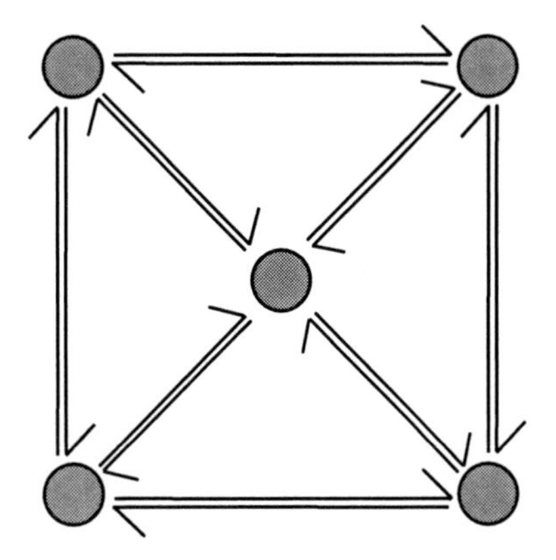

centric network structures are pivotal, with such centers usually referred to as "hubs." Ensuring convenient connectivity requires two key criteria: the frequency of connections offered over a particular time period (a day or a week) and the time it takes to safely connect. Given the tremendous commercial impact of hub connectivity, hubbed airlines take great care to optimize the timing of aircraft not only in terms of production efficiency, but also for network connectivity—a key aspect of product quality. The list of competitive connections within a network is a set of carefully managed paths through a few central nodes or hubs.

In the United States, airlines required the blessing of governmental "designation" before they could serve a route until 1978, when the Carter administration paved the way for full deregulation. As a result, almost all major US airlines quickly moved to replace traditional point-to-point (P2P) networks by sophisticated and connectivity-driven hub-and-spoke systems. In Asia, for instance, we still observe many highly regulated markets and governments designating national traffic routes and fares, with network structures far less advanced than their counterparts in highly competitive, deregulated environments.

Why were US airlines in the early 1980s—and European airlines soon thereafter—so eager to jump onboard hub-and-spoke systems, even though this meant asking passengers to connect on markets they traditionally flew non-stop? In other words, why did the economics of hubbing look so tempting?

Formula (2) calculates the number of routes needed in a P2P network (see Fig. 10) to connect all destinations with each other:

$$Q = n \times (n - 1) \tag{2}$$

where Q number of directional routes, n number of airports served, including hubs.

Sometimes, the formulas $Q = \frac{n \times (n-1)}{2}$ or $Q = \frac{n \times (n+1)}{2}$ can be found in the literature.

- Dividing by 2, these formulas consider routes to be bidirectional. Routes, however, are directional in both commercial and operational terms. The background of the somewhat careless use of the division by 2 is rooted in network

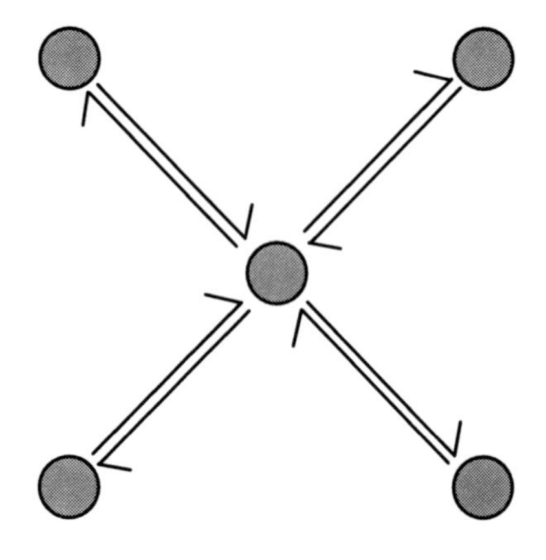

Fig. 11 In a hubbed system, five nodes can be connected, including transfer connections, by eight routes

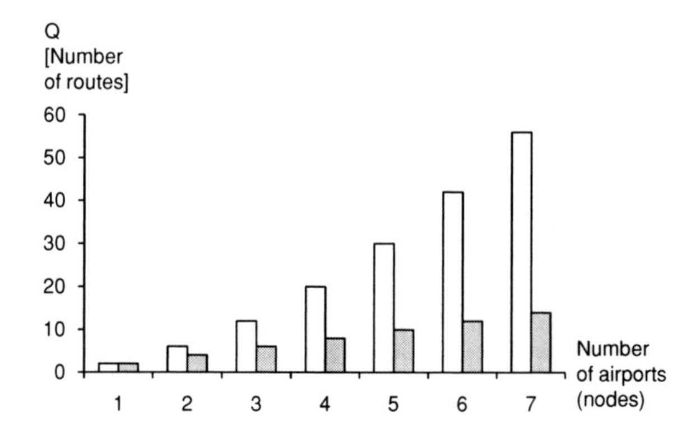

Fig. 12 Point-to-point versus hubbing. The number of routes needed in a point-to-point network [*white bars*: $Q = n(n - 1)$] grows faster than the connections of a hubbed network (*gray bars*: $Q = 2(n - 1)$]

topology theory where the number of edges in a network of n nodes is correctly calculated by dividing by 2, because a graphical "edge" is non-directional. Since flights or O&Ds are invariably directional, the "counting" of bidirectional edges is inappropriate in airline networks.

- The term $n \times (n + 1)$ assumes that there are n airports to be connected with each other, plus a central hub (the "+1" term). While $n \times (n - 1)$ assumes the hub to be included in n, the term $n \times (n + 1)$ does not. Since hubs are airports as well, serving regular demand as an origin or destination, the term $n \times (n - 1)$ is appropriate in the context of aviation.

The number of routes needed for a perfect hub system (see Fig. 11), where a central hub connects with all spokes, results as

$$Q = 2(n - 1) \tag{3}$$

where Q = number of directional routes, and n = number of airports, including the central hub.

Figure 12 contrasts the number of routes needed to connect each airport with other airports as a function of a growing number of airports.

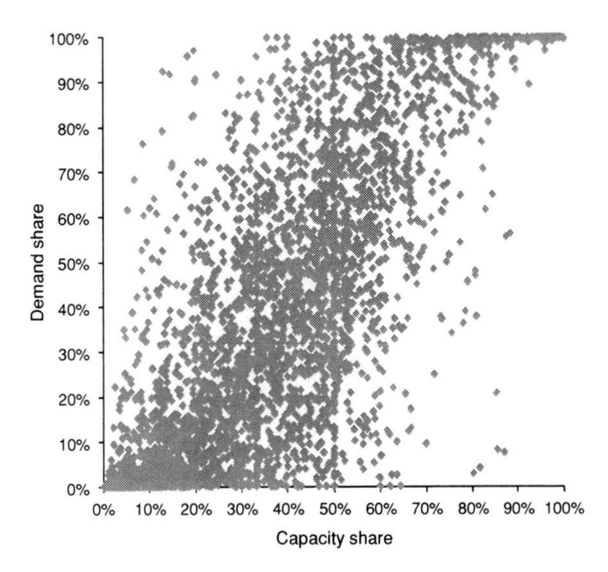

Fig. 13 The S-curve effect for 4,300 global and randomly selected O&Ds

As formulas (1) and (2) or the corresponding values in Fig. 12 make clear, the need for serviced routes develops disproportionately strong in a P2P system $(n \times (n - 1))$, whereas it remains linear in a hubbed network architecture $(2 \times (n - 1))$. The larger the network, the more efficiently it can be served through hubbing, assuming that demand per individual O&D is limited.

Hubs offer the chance for the dominating carrier to dominate the traffic at the respective airport. Such operational dominance can be expanded into commercial and other sorts of dominance, such as dominating brand appearance, control of corporate accounts, and price control. Often, this is referred to as the "S curve effect." This effect describes an under-proportionate share of demand as a function of a weak capacity share, whereby an over-proportionate share of capacity translates into an over-proportionate share of demand. The S curve effect must be differentiated for airports as origin, transfer hub, destination, or as per O&D or airport pair. At least at the level of O&Ds, however, the S curve effect is so weak (see Fig. 13) that it does not qualify as a strategic lever.

2.3.1 Connectivity: The Central Paradigm of Hub-and-Spoke Structures

Connectivity is defined as the ability to offer competitive connections. We will refer to competitive transfer connections as "hits." Hits must satisfy all of the following criteria:

- *Minimum connecting time* (MCT): a connection must occur within the minimum connection time, or MCT. The applicable MCTs are provided by the airports and

Arrive Station Code	Depart Station Code	Type of Connedction	Minimum Connection Time	One Wy Indicator	Connection Frequency	Arrive Carrier Code	Depart Carrier Code	Arrive Flight Number Range	Dpart Flight Number Range	Previous Station Code	Next Station Code	Effective Date	Discontinued Date
HKG	HKG	II	60	1	1234567	CX	CX		1727			20021111	29991231
HKG	HKG	II	60		1234567	CX	KA		1192			20080813	29991231
HKG	HKG	II	60		1234567	CX	KA		1196			20080813	29991231
HKG	HKG	II	60	1	1234567	CX	KA		1304		CAN	20080813	29991231
HKG	HKG	II	60		1234567	CX	KA		1306			20080813	29991231
HKG	HKG	II	60	1	1234567	CX	KA		1320		CAN	20080813	29991231
HKG	HKG	II	60	1	1234567	CX	KA		1383			20080813	29991231
HKG	HKG	II	60	1	1234567	CX	KA		1385			20080813	29991231
HKG	HKG	II	65	1	1234567	CX	LY	110	76			20090626	29991231
HKG	HKG	II	65	1	1234567	CZ	CO					20060412	29991231
HKG	HKG	II	70	1	1234567	CA	CI			PEK	TWN	20090812	29991231
HKG	HKG	II	70	1	1234567	CA	AE			TSN		20090812	29991231
HKG	HKG	II	70	1	1234567	CA	CI			TSN		20090812	29991231
HKG	HKG	II	70	1	1234567	CA	CI			TSN	TWN	20090812	29991231

Fig. 14 Example of an MCT table for HKG (September 2009, Innovata)

are updated monthly. Usually, each airport publishes a list of standard MCTs along with many exceptions. MCTs depend, for instance, on the domestic or international dimension of the connection at hand. MCT tables therefore differentiate MCTs for domestic-to-domestic (DD), domestic-to-international (DI), international-to-domestic (ID), or international-to-international (II) connections. Since connections with at least one international segment usually require customs procedures, they typically demand a longer MCT than a connection between two short-haul domestic flights. Other criteria in MCT tables relate to special flight numbers, airlines, origins, destinations, arrival or departure terminals, or validity periods. By publishing the MCT, the respective airport guarantees that all connections compliant with the rules as set out in the MCT table are feasible, for both passengers and baggage. Since many airports publish large numbers of MCT exceptions, full compliance with all MCT exceptions is indispensable when assessing connectivity. Figure 14 shows an excerpt from a typical MCT table.

Because small airports tend to have significantly shorter MCTs, they can provide faster transfer connections (see Sect. 3.7 in Chap. 3). Large airports, on the other hand, are usually much more constrained and can only offer relatively slow connections. (In Fig. 49, we present quantitative evidence for this observation.) Thus, airlines operating from smaller airports, or "hublets," can offer faster connections than those operating in larger hubs.

- *Detour*: the detour imposed by the connection must be sufficiently convenient. The detour is defined as the ratio of the total of the distances of the inbound and outbound legs, over the greater circle distance between the origin of the first leg and the destination of the second (or third in case of three segment connections).

As booking data reveal, passengers accept significantly larger detour factors for short elapsed times; while for long-haul flights, acceptable detours appear much tighter. Detour factors permit some backtracking, which is particularly relevant for long-haul connections. Reasonable detour factors for long haul are in the order of 1.2, and more generous detour factors are appropriate for short/medium haul (from 1.35 to 2.5) connections.

- *Bi-directionality*: the connection must be offered in both directions at least once per week (bi-directionality). The reason for this criterion is that it is difficult to sell a transfer connection in one direction without being able to offer a return connection—non-stop or transfer.
- *Traffic right restrictions*: as defined by IATA and as documented in the Standard Schedules Information Manual (SSIM) published by IATA (SSIM 2008), traffic right restrictions must be satisfied.

Hits must be sufficiently fast to qualify as competitive, and those that are too slow must be discarded. Hence, the question arises: how is "sufficiently fast" or "too slow" determined? There are two prevailing approaches for quantitatively assessing connectivity: One is based on a fixed time window, and the other on a flexible time window.

2.3.1.1 Fixed Maximum Connecting Time Windows

In most of the literature on this subject, the number of connections is defined by a fixed time window, referred to as "maximum connecting time," or abbreviated as MaxCT (see Fig. 15). The idea behind this concept is that meaningful connections

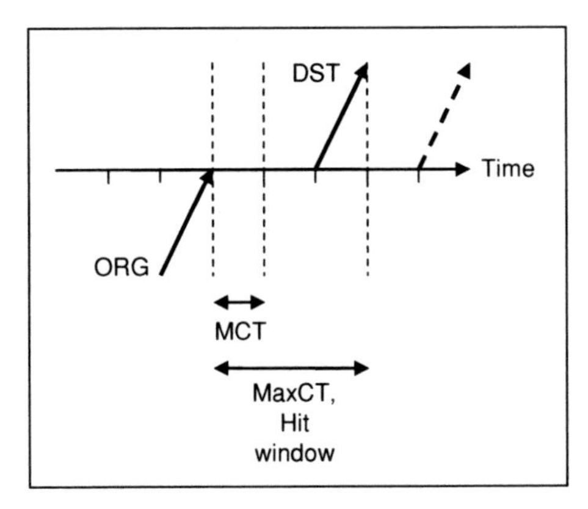

Fig. 15 Hit definition based on a fixed hit window. Connections must depart after completion of the applicable MCT, and before completion of a pre-defined time window, called maximum connecting time, or MaxCT. Additional criteria, such as detour factors, the requirement of an offer in both directions of the underlying O&D, and the absence of a faster non-stop connection, may also come into play

must happen after MCT and within the timeframe defined by MaxCT. Many different proposals exist as parameters for hit windows: Doganis and Dennis (1989) propose a standard hit window of 90 min for all types of connections. Bootsma (1997) proposes 180 min for connections between connecting continental flights, 300 min if one intercontinental flight is involved, and 720 min for connections between two intercontinental flights. Danesi (2006) suggests a differentiated set of values ranging from 90 to 180 min.

However, all approaches trying to define fixed MaxCTs have a common and significant disadvantage. A fixed hit window is of limited value for comparative purposes. A MaxCT that is competitively relevant for short/medium-haul to short/medium-haul connections in Europe is likely to be far too aggressive for many airports, even large ones, in Asia or South America. If a MaxCT of 120 min may be appropriate for Europe (medium–medium haul), at least twice as long a time window is needed to meet common and competitive transfer times of the same type at large hubs in Asia. On the other hand, if the wide-open MaxCT time windows of Asian hubs were applied to connectivity-optimized hub structures in Europe or the United States, the MaxCT would count far too many connections as being competitive.

2.3.1.2 Auto-Adaptive Hit Windows

An auto-adaptive time window takes the globally fastest possible elapsed time on a given O&D (regardless of the point of transfer), including non-stops, and at a given time period (between the time of departure and time of arrival). It takes the elapsed time of this connection as the reference, no matter how fast or slow this elapsed time may be in absolute terms (Burghouwt 2007; Burghouwt and Redoni 2009; Malighetti et al. 2008; Paleari et al. 2009). Note that for an auto-adaptive hit window, elapsed time is the reference; whereas for fixed time windows, it is the connecting time only. Any hit significantly slower (60 min in total) than the reference hit is discarded. At least conceptually, the time window to find the fastest possible connection remains open for an unlimited time period (see Fig. 16). As a result, the duration of a hit window becomes auto-adaptive toward the respective competitive environment of hits on each individual O&D worldwide. If the fastest transfer connection on a given O&D takes four hours to connect—significantly extending the resulting total elapsed time for the journey—then this elapsed time, though slow in absolute terms, is the only relevant reference for determining the competitiveness of connections on this specific O&D. For benchmarking purposes, the application of an auto-adaptive hit window or MaxCT is the only way to ensure direct comparability of the connectivity of hubs located in diverse markets. However, due to its simplicity in concept and implementation, a fixed MaxCT is preferable when focusing on structuring or evaluating schedule scenarios at a particular hub.

Connections that satisfy all of these criteria—MCT, detour, bi-directionality, granted traffic rights, and fixed or auto-adaptive hit window—are referred to as

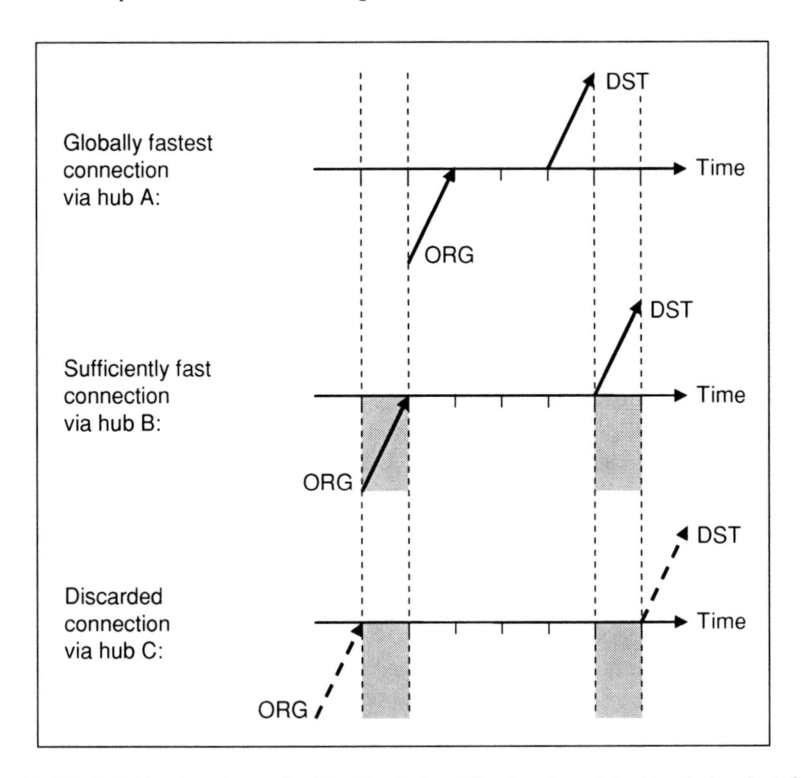

Fig. 16 Hit definition based on a flexible hit window. The duration of the hit window is defined by the duration of the globally fastest hit on the respective O&D at the time span given between scheduled time of departure (STD) of the inbound and scheduled time of arrival (STA) of the corresponding outbound flight, plus some buffer added. The auto-adaptive hit window then discards any hit slower than the timeframe set by the fastest hit and its buffer

"competitive hits" or "hits." In the framework of auto-adaptive hit windows, the consideration of detours is less relevant since an excessive detour invariably creates a slow elapsed time on a given O&D. This, in turn, makes it likely that such a connection is dominated by a less detoured connection. For the purpose of computational efficiency, however, detour factors offer advantages in the context of auto-adaptive hit windows.

2.3.2 *Connectivity and Codeshares: Camouflage or Mimicry?*

Frequently, airlines operating a particular flight offer the opportunity to one or more airlines to sell this flight under their own flight code (airline code plus flight number), falsely representing to the passenger that this flight is operated by the other airline. This mechanism is called "codesharing." The purpose of codesharing is to improve sales due to two factors: (1) A flight carrying a familiar airline code sells better than a flight carrying an unfamiliar code; and (2), in GDS displays, transfer connections based on flights carrying the same airline code are displayed better than others, even if one of the two codes is based on a codeshare.

Should codeshares be considered when evaluating connectivity? If so, how? When operationally designing a hub schedule, the scheduler can only plan for the flights under the control of their airline (or online hits, where both contributing flight legs are operated by the same airline). This would suggest that for planning purposes, only operated flights may be considered. However, if the arrival or departure times of long-haul codeshares are fixed and the scheduler must plan around (de-)feeder flights, codeshares do play a central role in planning. The same applies for flights protected by antitrust-immunity (ATI). For marketing purposes, full codeshared connections are obviously beneficial. Interline connections are between flights where both legs are operated by a different airline and where no codeshare exists for these connections.

Depending on the nature of the operating or codeshared proportions, there are four distinct levels of connections:

- *Operating online*: both legs of the connection are operated by the same airline.
- *Partial codeshare*: one leg carries a codeshare, which is pivotal in establishing the respective connection.
- *Full codeshare*: both legs are codeshared only, and neither flight is operated by the airline represented by the codeshare code.
- *Interline*: the connection can only be established based on operating and/or shared codes from different airlines.

When defining hits, these levels of sharing or not sharing codes play a key role. For instance, should an operating online hit qualify as a hit if there is a faster interline connection? The applicable rules will depend on the case at hand. However, as a general rule:

- Higher levels in the connections ranking above beat lower levels.
- Faster online connections beat slower connections.

An effective way to implement such rules is by adding time penalties to the elapsed time of lower-ranking codeshares, with multiple penalties accumulating.

2.3.3 Assessing Connectivity via Connection Builders

Connection-building computer programs (CBs) exist in many variations, depending on the purpose. The Computer Reservation System (CRS) represents the most comprehensive implementation of building connections. A CRS is legally prevented from weighing connections ("biased display"), with the notable exception that certain rules apply to sort connections for display. CRSs tend to build and offer many unreasonable connections, requiring excessive detours despite the availability of more convenient connections. Typically, CRSs prioritize connections by sorting according to total elapsed travel time. To support schedule development or the competitive evaluation of schedules or schedule scenarios, CBs must apply tight definitions of competitiveness; with loose definitions, competitive connections cannot be differentiated from poor connections. A common weakness of many CBs is that they do not fully respect MCT exceptions. While often considered an "average out" shortcut, CBs can lead to severely misleading results for hubs with complex MCT rules, such as Paris CDG. Moreover, the definition of MaxCT (width of time window, parameters of fixed and/or auto-adaptive hit windows) varies widely between the various implementations of CBs. Therefore, the exact assessment of feasible hits in complex banking designs quickly becomes intricate and requires tool support (Jost 2009).

2.3.4 Evaluating Schedules with QSIs and Market Share Models

To estimate the likely market shares of a particular flight on a given O&D, one must estimate the "utility" of that flight or connection for typical passengers. One proven way is to mirror passenger decision making, based on the empirical data of real-life passengers. Let us assume a passenger wants to fly from Manchester/GB (MAN) to Istanbul/Turkey (IST). The hypothetical passenger goes to a travel agency in Manchester and asks for an option to fly from MAN to IST. The computer system in the travel agency will display a few dozen potential connections, including non-stop and transfer connections via LON, AMS, CDG, FRA, and other transfer points. What kind of evaluation criteria will the passenger apply to identify the most attractive connection? The passenger will probably compare the total travel time, the reputation of the respective airline, departure and arrival times, and the applicable airfare. Depending on the relative importance of these and other criteria, the passenger will finally opt for one particular connection. A market share model tries to emulate this rationale of decision making by quantifying all kinds of relevant quality criteria, and then applies appropriate weighting factors to each criterion. Three model families prevail:

Logit assesses the "utility" of a product or service along a series of criteria if compared against competing alternatives. "The probability of [a particular] purchase is its share of the utilities after exponentiation" (Lilien 1992; Coldren and Koppelman 2003). To apply the logit modeling to the problem of likely market shares of competing aviation itineraries, one must follow the four-step sequence of decision making used by the hypothetical passenger. The fourth step is to examine the resulting impact on likely market share. Figure 17 shows how logit models function in network analysis.

Neural networks are best understood as black boxes: they take large samples of "real" booking data to "learn" the interrelations between variables, and then apply what they have learned to predict passenger preferences for unknown or scenario markets. Neural networks yield results that in many cases are superior to logit-based models (Grosche and Rothlauf 2007). By definition, however, they do not

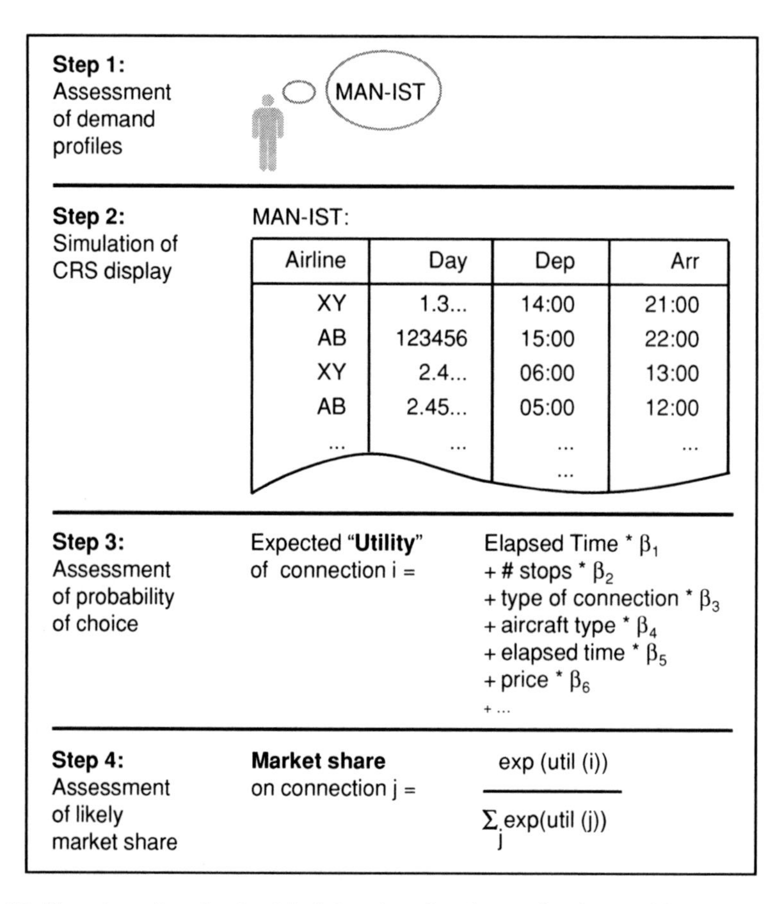

Fig. 17 The schematic rationale of logit-based market-share estimation models

provide insight into the rules that they have learned and applied. Many users of neural network based market share models shy away from accepting the outcome without understanding the reason why the model reached this result. Logit models, in contrast, offer reasonable results with maximum transparency of the computational process.

Quality of service indices follow the same rationale of mirroring passenger decision making as logit models do. These models calculate a "quality of service index" (QSI) for each connection opportunity, and take the share of the QSI score of one particular connection opportunity out of all connection opportunities on the selected O&D as a proxy for the likely resulting market share. Logit models are a variant of QSI models in that the logit model assumes a particular relationship between the factors driving the decision making.

The various QSI implementations differ in calculating what and how many criteria are used, how the relative weighting factors are determined, and how individual weighted factors are combined into one overall QSI.

2.3.5 Spill and Recapture

When planning the market performance of a network, available aircraft capacity is considered in the context of the demand on each route. With the aid of complex mathematical models, passenger demand volumes are estimated per flight segment, including behind and beyond transfer traffic. Due to operational constraints, planners frequently must assign aircraft to a route that is too small to accommodate the expected demand for that route. As a result, otherwise potential passengers are "spilled." In large multi-hub networks, however, chances are that such "spilled" passengers will opt for a later flight or different itinerary of the same O&D served by the same airline. This effect is called "recapture" of spilled traffic. State-of-the-art planning tools take into account the recaptured traffic when estimating available demand per route.

2.4 Point-to-Point: The Answer to Hub-and-Spoke?

In highly regulated aviation markets, one finds a clear dominance of decentralized network structures with P2P services almost exclusively. An immediate effect of deregulation in the United States, or of liberalization in Europe, was that network architectures quickly shifted into significantly more centralized, or hubbed, topologies (see Fig. 18).

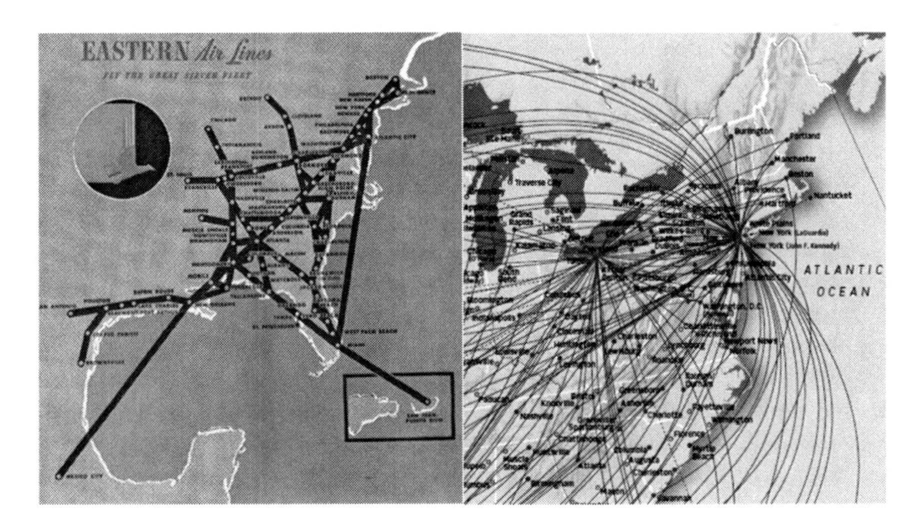

Fig. 18 Regulated versus deregulated network structures. *Left* the decentralized network structures of a prominent US pre-deregulation airline (United States Accountability Office, GAO 2006). *Right* example of a US hub-and-spoke network in 2009 (Antenna Audio Inc.)

Hub-and-spoke offered the opportunity to drastically expand the network scope—with the same fleet size inherited from former regulated times—particularly into markets too small for non-stop services, or to maintain the original network scope but with a much smaller fleet. The economics appeared simple and attractive: If an airline already serves flights from HAM to FRA and from FRA to FLR (thus covering most costs except the passenger variable cost for sales commission, peanuts, or a soft drink), passengers flying from HAM to FLR and connecting in FRA would come at marginal costs.

One caveat: when airlines observed the stunning success and strong growth of connectivity-based hub-and-spoke systems, many ordered larger aircraft to accommodate the apparently growing segment of transfer passengers. Many airlines overlooked the fact, however, that the transfer traffic only comes at marginal costs when the underlying capacity is aligned to P2P demand. When an airline opts for a larger aircraft offering more seats for transferring passengers, this capacity expansion comes at full, not marginal, costs. Many airlines had to cut back over-dimensioned aircraft when they found out about these fundamental economics. The aircraft manufacturers responded with new aircraft designs, particularly for long haul, offering relatively small capacity while maintaining high-distance reach. With these types of aircraft, airlines can afford to offer long-haul services even with limited volumes of transfer traffic, effectively keeping the full costs of transferring services at bay. At the same time, "super jumbos" like the A380 are introduced to cut down costs of high-volume, long-haul trunk routes.

In addition to the marginal cost issue, connecting passengers require many costly extra services that further complicate costs. While some of these costs are related to the spiked or "wave-like" schedule structures discussed later in Chap. 3, it is appropriate to now summarize the respective cost drivers. Here is a partial list of examples:

- Reduced asset utilization:
  - Non-optimum utilization of ground resources: Transfer-driven hub structures frequently cluster traffic into banks of high-traffic activities, typically separated by periods of comparatively low activity. All ground facilities and procedures must be tuned to maximum activity during such peak times, resulting in underutilization of resources during the periods of low activity. This applies to resources from counter positions to push-back vehicles.
  - Non-optimum utilization of aircraft and flight crew: In any such banked or "waved" system, many aircraft must wait at the preceding outstation or at the hub itself to match the timing requirements of a given inbound bank. Such idle time of the most expensive resource of an airline—aircraft plus flight crew—drains a lot of money.
- Enhanced commercial complexity:
  - Special complex algorithms in inventory or pricing systems.
  - Complex market research and competitive analysis.

- Heightened operational complexity:
  - Rapid transfer of passengers and their baggage to the departing gate also requires the airport to invest in high-performance baggage sorting and transportation devices.
  - Extra security measures and security infrastructure (note the required strict separation of Schengen- and non-Schengen passengers in Europe, requiring multimillion Euro investments from most international airports in Europe).
  - Waved patterns are vulnerable to delays: In a highly waved, spiked system, any delayed inbound flight easily creates delays or missed connections. Considering the connection-timing profile shown in Fig. 19 with its sharp rise of connections right after completion of the MCT, any delayed inbound flight would affect a few, or many, connections. Delayed inbound flights create operational frictions with regard to rotational plans for aircraft and crew, as well as for gate, runway, and air space capacities. Compare the connection timing profiles of two major US airlines at their respective prime hubs (see Fig. 19). Airline A starts building up connections as early as 25 min after arrival. The highest frequency of connections (referred to as "hits," see Sect. 2.3.1), however, is reached as late as 65 min after arrival. Airline B, in contrast, starts building up hits at 30 min and reaches its maximum only 15 min later. While the "A" pattern (solid line) is likely to offer higher scores of punctuality, the "B" structure (dotted line) is likely to appear more attractive on CRS displays.

In essence, connecting passengers can create complex costs that could severely curtail the efficiency advantages of hubbed network architectures.

The strategy and implementation of hubbed networks, while aiming at optimum coverage of as many O&Ds as possible, created its own worst enemy: LCCs. While hubbed network carriers assume that transfer O&Ds are too thin to permit direct services and cannot be served at lower costs except through hubbing, LCCs challenged both convictions. The LCC business model assumes that eliminating all the complex costs mentioned above will result in significantly lower costs, a more aggressive price point, and a boost in demand. Hence, LCCs attack hubbed

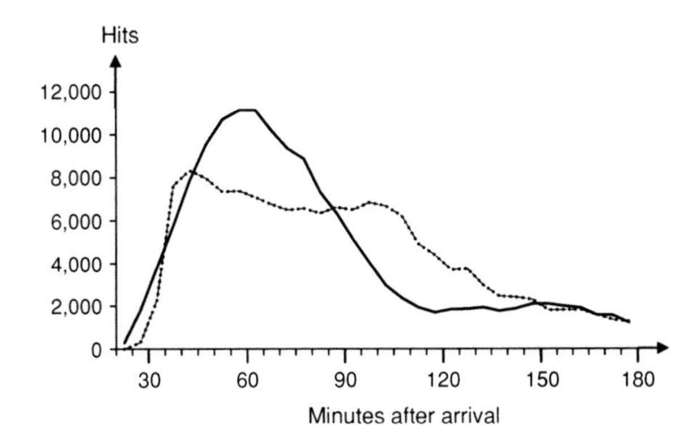

Fig. 19 Connection timing profile of two major US airlines at their respective home bases

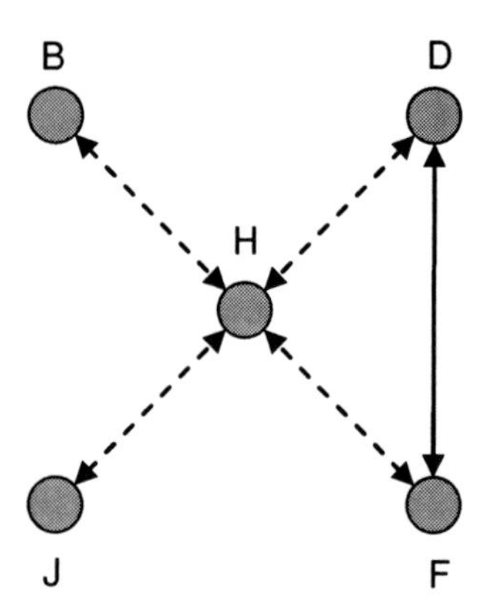

Fig. 20 LCCs (*solid line*) attack hub systems where they are most vulnerable—thin O&Ds which hub networks only serve as transfer traffic (*dotted lines*)

networks where they are most vulnerable: by serving O&Ds non-stop, which otherwise require transfers in hubbed systems (see Fig. 20). By doing so, LCCs not only grow their own business, but also simultaneously erode the demand base and the price point of their hubbed counterparts.

To remain competitive against non-stop services, hubbed carriers must offer fast connections. A slow transfer connection adds two inconveniences: the need to connect and the waste of time. The need to offer fast connections explains, at least partially, why hubbed airlines invest so heavily in sophisticated and operationally demanding high-connectivity network structures. In addition, reservation systems sort alternative connections in ascending order of total elapsed time, prioritizing faster over slower connections.

How can non-stop services compete against hubbed connections in terms of cost structure, given the enormous economies of scale inherent to hubbed systems? Frequently, successful non-stop services are offered by LCCs in direct competition with hubbed carriers.

LCCs follow different economies of scale. While network carriers leverage the massive number of city pairs they can serve at high frequency and thus generate their economies of scale, LCCs leverage the standardization of their production platform (a uniform fleet) to achieve similar or superior unit cost advantages and to permit simpler and more efficient procedures (TAT). Furthermore, LCCs exploit any other opportunity to reduce production costs. Passengers opting for the non-stop service of an LCC benefit from the convenience of a non-stop service and a more aggressive fare, but must accept the lack of differentiated service concepts on the ground and on the flight due to LCC standardization. Passengers preferring the offer of a network carrier on the same O&D may need to accept the inconvenience of a transfer, but in return have a much broader choice of connections as well as differentiated and convenient services. Increasingly, hubbed airlines and hub airports are cooperating to counter the inconvenience of connections.

Both business models, network as well as LCC carriers, are built upon economies of scale. However, both models rely on different levers to achieve such economies of scale. For the passenger, the complementary nature of both business models offers choices beyond the capability of each business model on its own. Two observations are interesting in this context:

- LCCs serve many transfer passengers and build their schedules accordingly, converging toward the business model of network carriers.
- Network carriers and hub airports, in turn, emphasize the need of high-convenience services at the point of transfer to counterbalance the intrinsic inconvenience of transfer.

2.4.1 Stuck in Between Hubs and Spokes? On Hublets

Hublets are airports with a high share of connecting traffic but built upon relatively small local demand. Thus, hublets strongly leverage—and in some cases over-leverage—limited local demand. Vienna (VIE), Austria might serve as an example of such a hublet: Austrian Airlines (OS) serves VIE airport as its homebase, serving a local demand originating in VIE in the order of 12 Mio passengers p.a., 54 destinations, and a fleet of about 27 aircraft (all 2007 figures). With this framework, OS has achieved a transfer rate of about 60% for its operations in VIE. The Vienna hublet builds upon a small catchment area, but ranks high in terms of relative connectivity if compared to the largest hubs in Europe. Without its transferring passengers, OS could not serve as many destinations and frequencies. The high rate of transfer traffic enables OS to offer a wide network of destinations and dense frequencies to local passengers in Vienna. Given the prime importance of such a broad offering of flights for the community and economy in and around Vienna, a high-performing transfer system is critical for this market. For a hub like London Heathrow (LHR), transfer traffic is a welcomed windfall profit; for hublets like VIE it is a matter of life or death. The strategic risk for hublets, however, is fragmentation. If too many and too thin O&Ds are served—and if too many small sales organizations and corporate accounts must be maintained—costs are likely to explode and exceed feasible revenues. This is particularly true if the hublet network significantly overlaps with competing hubs or hublets. The continuous decline of the PIT hublet in the north-east US over the last decade, with its limited catchment area and nearly complete overlap with surrounding hubs, depicts the vulnerability of hublet strategies.

Chapter 3
Designing Connectivity-Driven Network and Hub Structures

Abstract The most effective driver of connectivity is the number of flight movements at a given hub, as the number of feasible hits greatly rises when compared to the number of underlying flight movements. Contrary to popular belief, the number of banks (a temporal cluster of inbound and outbound flights) correlates conversely with connectivity: the fewer the better. Bank structures, as well as geographical, operational, infrastructural, and regulatory issues, deeply affect connectivity. Airlines have implemented or abolished structural variants, including rolling, random, or continuous hubbing, with varied success. In this chapter, we review the economics of these structural designs.

For quite some time, network managers have been using the terms "continuous," "rolling," "random," or "de-peaking" in their proposals for budget changes or career development moves. Depending on the network manager, however, these concepts may have different meanings.

Not everything that looks de-peaked is a copy of LCC concepts or a farewell to connectivity or hub-and-spoke. To comprehend the concepts, terminology, and methodology, we must first understand how the various drivers of connectivity lead to structural variants.

In determining connectivity, there are nine key factors or drivers:

1. Number of inbound and outbound flights
2. Temporal design of individual banks
3. Number of banks
4. Directionality
5. Rotational patterns
6. Airport infrastructure
7. Random connectivity
8. Minimum connecting time (MCT)
9. Internal structure of banks.

P. Goedeking, *Networks in Aviation*, DOI: 10.1007/978-3-642-13764-8_3,

3.1 Connectivity Driver #1: Number of Inbound and Outbound Flights

Let us assume that 100 inbound flights arrive directly from different origins in the West within 1 h. Starting 1 h later, another 100 flights depart to different destinations in the East within 1 h. Assuming an MCT of 60 min, all inbound flights could connect with all outbound flights. The resulting number of feasible connections is the product of the number of inbound and the number of outbound flights. We get

$$\text{hits} = \text{inb} * \text{outb} \tag{4}$$

where inb, number of inbound flights; outb, number of outbound flights; and hits, number of hits.

In the simplistic case of 10 inbound and 10 outbound flights, all within a single and well-timed bank and any other potential constraints disregarded, the airport would offer $10 \times 10 = 100$ connections, which is equivalent to the theoretical maximum. An airport offering 10 times as many inbounds and outbounds would not generate 10 times as many connections, but rather 100 times as many ($100 \times 100 = 10{,}000$). This clearly demonstrates the enormous economies of scale inherent to hubs.

3.2 Connectivity Driver #2: Temporal Design of Individual Banks

A bank is a temporal cluster of inbound and outbound flights, whereby the inbound flights all arrive within a relatively short and limited time period, and the corresponding outbound flights depart within a short time period once most or all of the inbound flights have arrived. This way, the likelihood of creating many fast connections is optimized if compared to a random distribution of flights. The cluster of inbound flights is called an inbound bank, and the outbound cluster an outbound bank. A bank is the combined entity of an inbound bank and its corresponding outbound bank. An inbound bank is also called a "feeder bank," and an outbound bank is a "de-feeder bank," referring to the feed/de-feed proportions of corresponding banks. Inbound flights within a bank may originate from similar or varied directions. The term "complex" is sometimes used as a synonym for "bank." Figure 21 shows the schematic of a system with one inbound bank and one outbound bank. For the sake of simplicity, we will refer to fixed maximum connecting times (MaxCT) (see Sect. 2.3.1 in Chap. 2).

A wave is the combination of an inbound and a corresponding outbound bank, separated from other waves by periods of reduced activity. The term "wave" describes the wave-like overall pattern of movements in a system where a distinct inbound bank is followed by a distinct outbound bank, then again by a sequence of

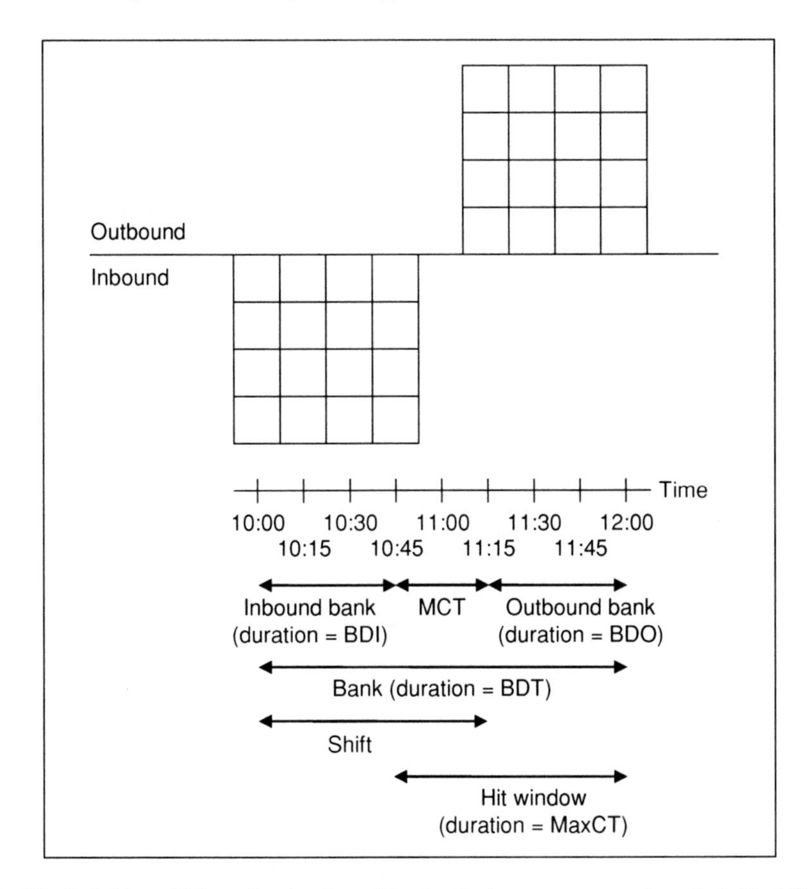

Fig. 21 Definition of inbound and outbound banks. Each *square* represents an individual flight

inbound-outbound banks, and so on.[5] The result is a continuous up-and-down pattern of flight activities, both inbound and outbound. Figure 21 is a schematic example of such a waved system, which is built by a series of nonoverlapping inbound/outbound banks. Note that "waved" and "banked" refer to distinct concepts. Fig. 55 in Chap. 4 shows an example of a clearly banked yet nonwaved hub structure. Most waved structures are built upon banks, but banked systems are not necessarily waved. [See Sect. 4.1.7 in Chap. 4 for an in-depth discussion of banked but nonwaved ("rolling") hubs.]

Figure 21 summarizes the definitions of the various time proportions of banks needed for connectivity calculations. BDI refers to the actual duration of the inbound bank, measured from the arrival time of the first flight within the given

---

[5] Bootsma (1997) defined a wave as a "complex of incoming and outgoing flights, structured such that all incoming flights connect to all outgoing flights." This definition assumes identity between banks and waves. As shown in Sect. 4.1.6 in Chap. 4, connectivity-driven banks can be arranged in nonwaved structures. Hence, a clear distinction between banks and waves is required.

inbound bank to the arrival time of the last inbound within this bank. BDO refers to the actual duration of the outbound wave, measured from the time of departure of the first outbound of the outbound bank until the last outbound within this bank. BDT measures the duration from the onset of the inbound bank to the completion of the outbound bank. "Shift" refers to the temporal displacement of the beginning of the outbound wave against the beginning of the corresponding inbound wave. The "hit window" describes the time available for an inbound flight to potentially connect with outbound flights (with MCT being the initial part of the hit window).

Under optimum connectivity circumstances, the last inbound flights in an inbound bank should be able to connect with the first outbound flights of the corresponding outbound bank. Otherwise, we would have to compromise on the number of feasible connections. So, the outbound wave must be shifted forward by at least the duration of the inbound bank plus the time span of MCT.

$$\text{Shift} = \text{BDI} + \text{MCT} \tag{5}$$

where shift, minimum time shift of outbound bank against inbound bank; BDI, duration of inbound bank.

In the example given in Fig. 21, the resulting minimum time shift is $45 + 30 = 75$ min.

The duration of the banks (BDI and BDO) must be such as to ensure that the earliest flights into an inbound bank can reach the last flights out of the corresponding outbound bank (particularly in the framework of a fixed hit window). MaxCT must be derived from competitive requirements, and the right BDI or BDO is the result of a given MaxCT. Hence:

$$\text{BDI} = \text{BDO} = \text{MaxCT} - \text{MCT} \tag{6}$$

Assuming a MaxCT of 75 min to be sufficiently competitive—and 30 min MCT—the optimum duration of the underlying banks is calculated as $75 - 30 = 45$ min.

Therefore, the pattern shown in Fig. 21 is the optimum structure for a wave length of 45 min and a shift of 75 min. It holds off the inbound and outbound waves by the time span needed for MCT (30 min), and does not exceed the MaxCT of 75 min to cover an inbound and outbound bank of 45 min each plus MCT. This schematically ideal example would yield 256 total hits (Fig. 22), as all

Fig. 22 Number of feasible hits for the ideal waved bank example in Fig. 21

Departure times					
	12:00	16	16	16	16
	11:45	16	16	16	16
	11:30	16	16	16	16
	11:15	16	16	16	16
		10:00	10:15	10:30	10:45
		Arrival times			

Σ = 256 hits

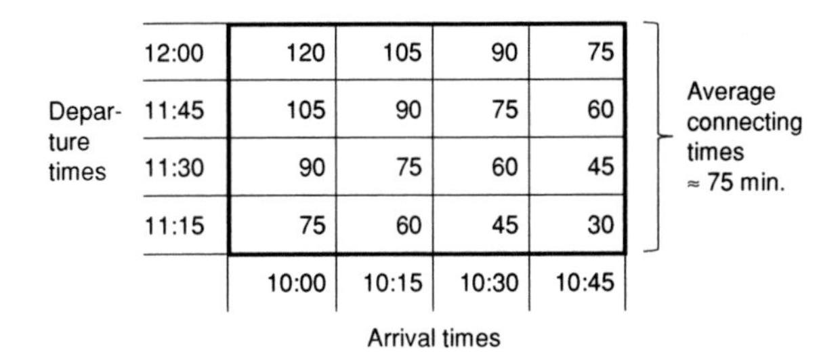

Departure times				
12:00	120	105	90	75
11:45	105	90	75	60
11:30	90	75	60	45
11:15	75	60	45	30
Arrival times	10:00	10:15	10:30	10:45

Average connecting times ≈ 75 min.

Fig. 23 Connecting times for the ideal waved bank example in Fig. 21

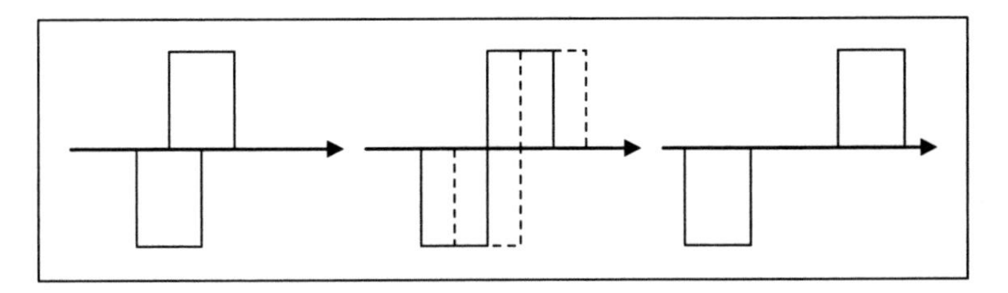

Fig. 24 Three types of overlap: Inbound–outbound (*left*), inbound–inbound or outbound–outbound (*middle*), and delayed outbound (*right*)

16 inbound flights can connect with all 16 outbound flights ($16 \times 16 = 256$). In Fig. 23, the connecting times for all hits are described, ranging from 30 min to 120 min, and averaging 75 min.

3.2.1 Bank Overlap

Bank structures strictly optimized toward connectivity are rarely found in the real world. Most often, requirements of aircraft and crew utilization lead to overlapping bank structures, either between inbound and outbound banks, or between subsequent inbound and outbound banks—or both. The following types of network overlap must be differentiated (see Fig. 24):

- *Inbound–outbound overlap*: overlap between an inbound and its corresponding subsequent outbound wave.
- *Inbound–inbound or outbound–outbound overlap*: overlap between successive inbound waves or between successive outbound waves.
- *Delayed outbound waves*, with the outbound wave being more distant from the feeding inbound bank than one would expect from the connectivity viewpoint.

3.2.2 Inbound–Outbound Overlap

How should an outbound wave follow an inbound wave? Should it overlap with its predecessor, with no or some delay? Let us first examine a conceptual case before

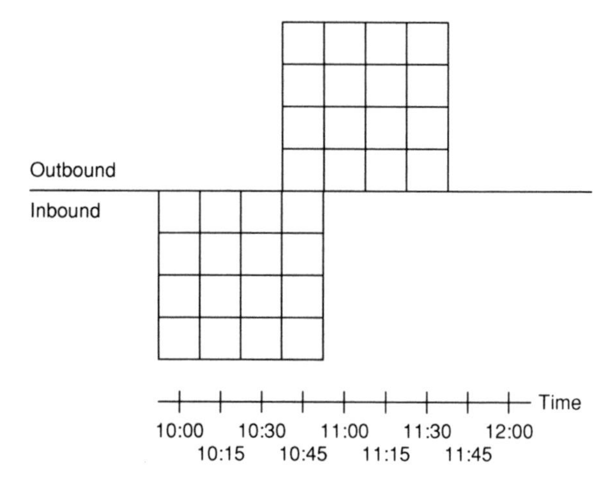

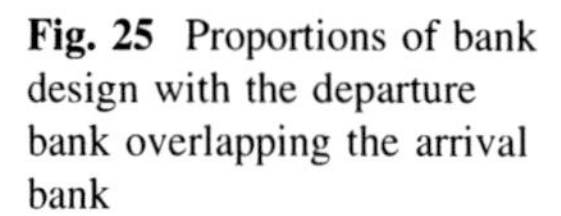
Fig. 25 Proportions of bank design with the departure bank overlapping the arrival bank

taking a closer look at actual cases of limited and extensive overlap between subsequent inbound and outbound banks.

In Fig. 25, we assume an MCT of 30 min; flights arriving at STA = 10:00 could connect at the earliest with flights departing at 10:00 + 30 = 10:30. Thus, all four flights arriving at 10:00 could connect with all 16 flights departing between 10:45 and 11:30 (assuming a sufficiently long MaxCT); this would result in $4 \times 16 = 64$ hits for flights arriving at 10:00. The fastest hits take 45 min to connect (10:00–10:45); the slowest hits take 90 min (10:00–11:30). The same hit rate applies to flights arriving at 10:15, as all four inbound flights can connect with all 16 outbound flights (each at 15-min faster connecting times). However, inbound flights arriving at 10:30 could only connect with flights departing at or after 11:00, resulting in 48 hits. The fastest hits take 30 min; the slowest hits take 60 min. This example would result in 208 hits and an average connecting time of about 53 min, ranging from 30 to 90 min (see Figs. 26, 27).

The loss of 48 hits (256 hits in the ideal case minus 208 hits in this case) is due to the close proximity of the outbound wave following the inbound wave. On the other hand, the average connecting time of the example shown in Fig. 21 would be as slow as 75 min (compared to 53 min in the overlapping example). Close proximity of inbound and outbound banks creates fewer but faster connections, while more separated banks tend to create more but slower connections.

As aircraft turnaround times are typically much shorter than passenger MCTs, bank designs that offer fast connecting times are more likely to offer higher aircraft utilization scores as well. If bank designs are optimized toward a maximum number of hits—shifting the outbound bank away from the inbound bank as suggested in Fig. 21—aircraft will have to wait for the outbound bank to commence long after operational minimum ground times would require. Poor aircraft utilization would be the result. In bank designs utilizing short bank durations, aircraft can turn around rapidly. In ideal cases, average passenger connecting time within a bank and aircraft turnaround times would be identical.

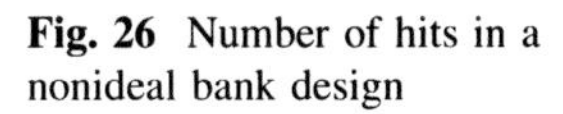

Fig. 26 Number of hits in a nonideal bank design

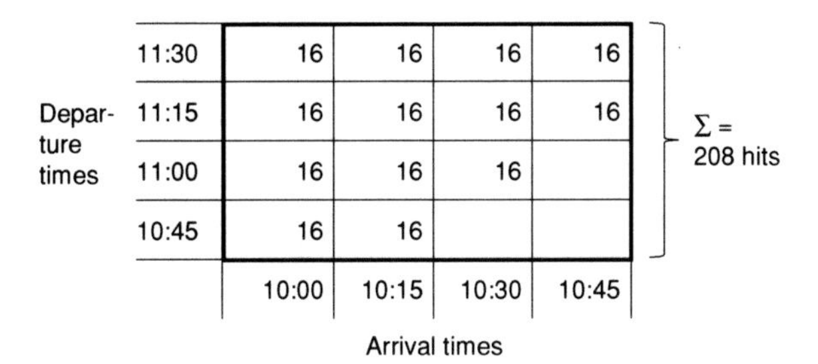

Departure times				
11:30	16	16	16	16
11:15	16	16	16	16
11:00	16	16	16	
10:45	16	16		
Arrival times	10:00	10:15	10:30	10:45

Σ = 208 hits

Fig. 27 Connection times for the nonideal bank design

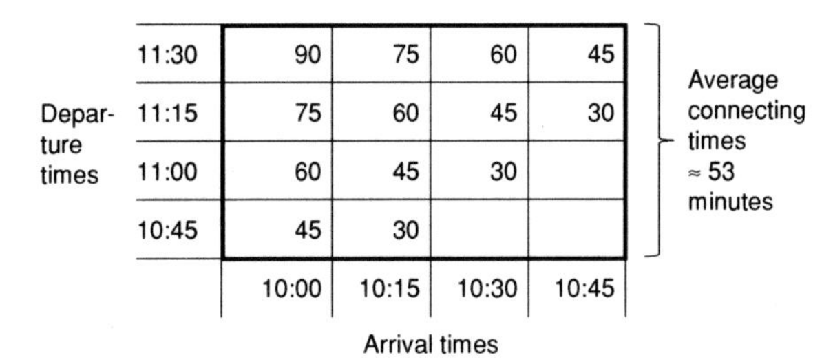

Departure times				
11:30	90	75	60	45
11:15	75	60	45	30
11:00	60	45	30	
10:45	45	30		
Arrival times	10:00	10:15	10:30	10:45

Average connecting times ≈ 53 minutes

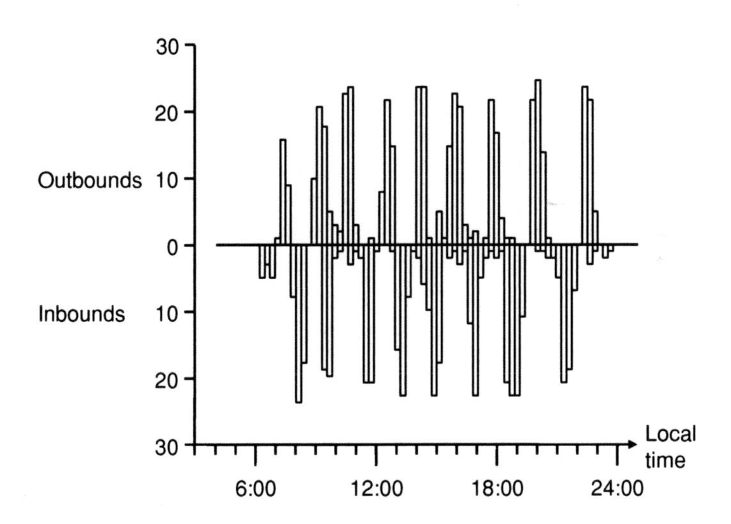

Fig. 28 Wave pattern of NW at DTW. Note the spiked structure of flight clusters, meant to optimize connectivity

NW at DTW (see Fig. 28) serves as a good example of how to combine high connectivity with fast TATs. Hence, the apparently conflicting objectives of a high number of hits, fast hits, and high aircraft utilization can only be combined by short, rapid bank structures. Because this requires a high number of banks, it can only be implemented at large hubs that provide the necessary volume of traffic. For smaller hubs or hublets, the number of hits, the speed of hits, and aircraft productivity become conflicting goals.

Quantifying the impact on the number of and speed of hits due to the overlap between an inbound and its corresponding outbound wave is complicated. The connectivity and productivity effect of bank overlap depends on the shape of the banks and other factors. While simulations may be the best way to quantify the impact, more practical approaches can serve the same purpose. If a perfect shift of MCT generates maximum connectivity (100%), and a full overlap eliminates any connectivity (0%), then the time course of overlap detriment moves from 100% down to 0%, more or less, within the limited bandwidth of a linear trend. The more overlap, the smaller the number of hits and the faster the remaining hits. We will refer to this effect as BOF or "bank overlap factor." BOF ranges from 0 (full overlap) to 1 (no overlap with ideal shift).

While Fig. 28 provides an example of a real waved hub structure, Fig. 29 offers a schematic example:

NW at DTW (see Fig. 30) provides an instructive example of individual banks succeeding with almost ideal delay.

The selected bank commences at about 19:36 and lasts until 20:59, making BDI (duration of the inbound bank) = 83 min. The outbound wave commences at 21:30 and lasts until 21:55. With MCT = 25, the optimum shift would be BDI + MCT or 83 + 25 = 108 min. Hence, the ideal starting point for the outbound wave, ensuring that the last inbound flight can safely connect to the first departures of the outbound wave, is 19:36 + 1:48 = 21:24. With the real starting point being scheduled at 21:30, NW has added an extra 6-min buffer. The rationale for this safe buffered timing is clear: This bank is the last major bank in DTW for passengers connecting West to East. If the passengers miss their connecting flight during this bank, they will have to stay overnight to catch the first flight in the morning. Hence, the safe timing is a means to increase connecting reliability and passenger satisfaction. Similar buffers can often be found in cases where the first inbound wave in the morning is mostly long haul and has to reliably connect with the first outbound bank. Those cases are often found in Europe.

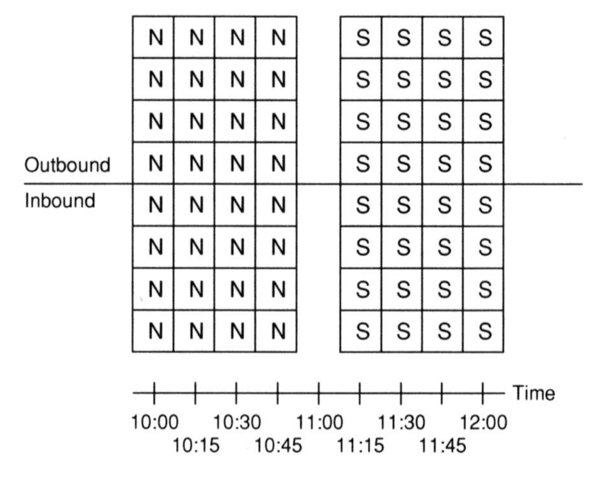

Fig. 29 Schematic example of a two-directional waved system, composed of flights to/from North (N) and South (S) only

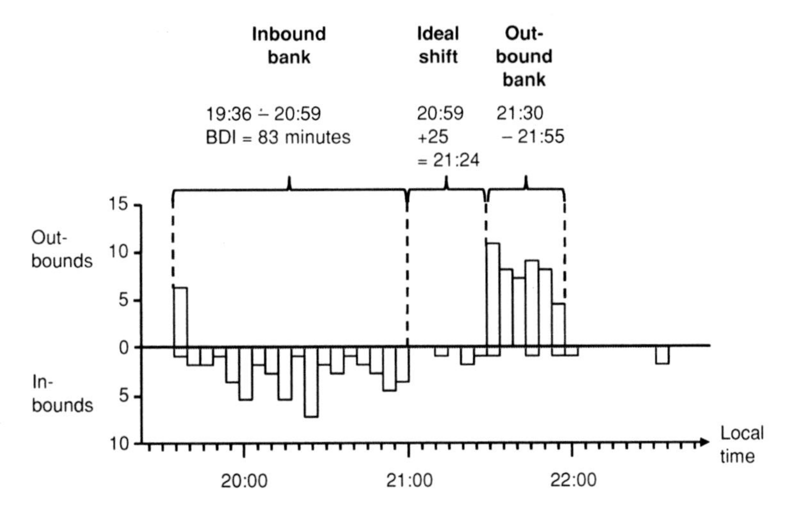

Fig. 30 Evening bank of NW at DTW, shown with a 5-min resolution

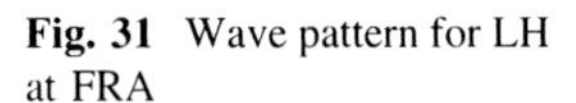
Fig. 31 Wave pattern for LH at FRA

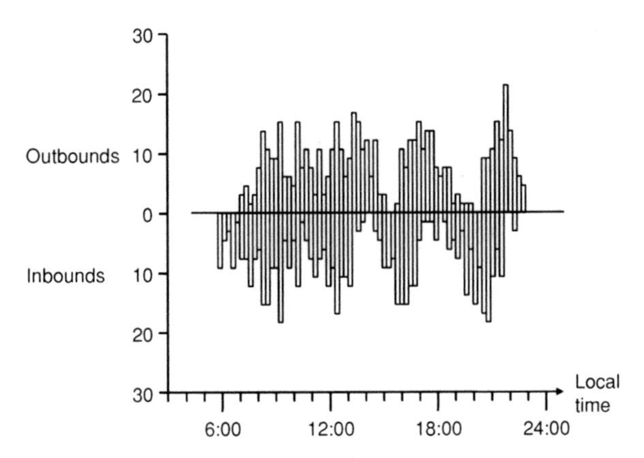

At its FRA hub, LH (see Fig. 31) exhibits an example of significant overlap between successive inbound–inbound or outbound–outbound banks:

Taking the evening bank as an example, the inbound wave lasts from 18:35 through 21:50 (BDI = 195 min), and the outbound wave lasts from 19:30 through 22:50 (BDO = 200 min). Applying formula (5) to compute the theoretically most apt displacement results in shift = 195 + 45 = 240 min. This contrasts sharply with the actual displacement of just 55 min (start of inbound 19:30 to the start of outbound 18:35 = 55 min). The reason for this deviation is threefold:

- On highly congested hubs around the globe, capacity constraints prevent schedule structures that one would choose in a less constrained environment. FRA has insufficient capacity to permit shorter hence more spiked bank patterns. Capacity limitations are usually the most severe factor in constraining connectivity-efficient bank designs. As a result, airlines operating at capacity-constrained hubs have to choose between either fast or many connections.

- FRA serves many connections involving long-haul flights and wide-body aircraft. Important long-haul flights depart toward the end of the last outbound wave in FRA to JNB, GRU, and BKK. Almost all inbounds of the last wave can safely connect to these late long-haul outbounds. Most short/medium-haul outbound flights during this bank mainly serve local rather than transfer traffic.
- There is a trade-off between overlap and the resulting number of connections. Consequently, LH may lose some hits due to significant bank overlap. On the other hand, LH gains competitive advantage through a significant acceleration of the hits it does obtain.

3.2.3 *Inbound–Inbound and Outbound–Outbound Overlap*

Conceptually, overlap between one outbound bank and its subsequent outbound bank (and the same for inbound) is a combination of waved and flat (rolling or continuous) hub patterns (see Fig. 55 in Chap. 4). In Sect. 4.1.6 in Chap. 4, we will discuss how flat hub patterns can serve as a means to structure the flow of traffic throughout the day. However, in this section, we will focus on the individual structural elements. Airlines might opt for such a combination for a variety of reasons: Infrastructural constraints may not permit spiked wave systems, or a flat pattern would be too costly in terms of volume or connection-time performance of hits.

3.3 Special Topic: Rapid Banking

In most banked hub structures, typical banks last for at least 2 h (inbound and outbound banks combined) (see Fig. 28). In some cases (see Fig. 31), banks may last as long as four hours. Such extended bank durations increase the probability of producing a large number of hits, but run the risk of most hits being fairly slow. Fig. 32 shows an example of a hub structure utilizing a much shorter bank duration

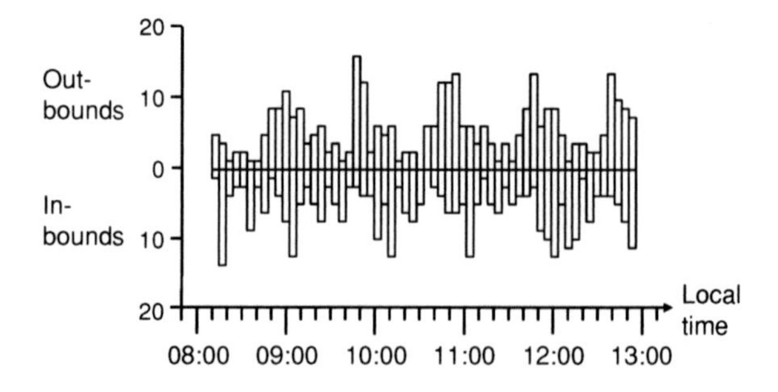

Fig. 32 Selected time slice of AA's DFW schedule in October 2005, shown in a 5-min grid

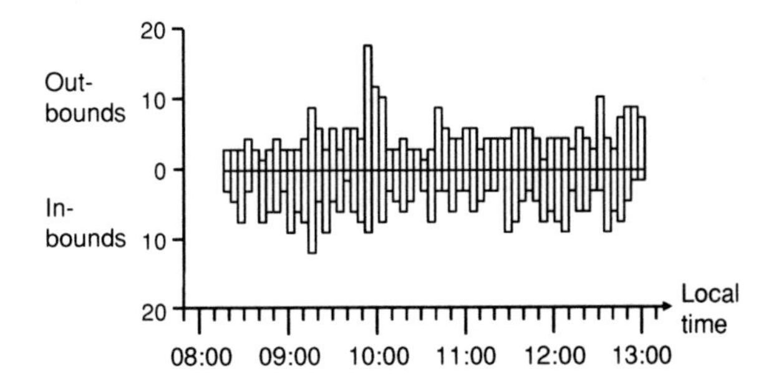

Fig. 33 The same time slice as in Fig. 32, but in September 2009

and bank repeat frequency. At its DFW hub in 2005, AA operated a system of banks of about 1-h duration, inbound and outbound banks combined, which resulted in about 14 short banks rapidly following one another. Such "rapid banks" attempt to combine the commercial goal of fast and many hits with the operational objective of minimizing the aircraft idle time needed when waiting for banks. Note that rapid banks of this kind still result in clearly waved, though less obvious, hub structures. Fig. 33 shows the same time period in AA's summer 2009 schedule, revealing that AA has since opted against this scheme.

3.4 Connectivity Driver #3: Number of Banks or Waves

The number of banks is closely related to the duration of individual banks. Long-duration banks, for instance, limit the overall number of feasible banks. However, the number of banks needed to cover an entire day is determined by more factors than just the duration of individual banks. Connectivity considerations, time-of-day coverage, and local demand profiles all exert significant impact on the optimum number of banks or waves.

As we examine the impact of a varying number of waves or banks on connectivity, we will assume optimum timing proportions within the individual banks, so inbound and outbound banks are separated by MCT. Because all calculations will equally apply to waves or banks, all considerations for banks can be applied to waves 1:1.

A single daily bank rarely meets the requirement to offer services at multiple times per day for both local and connecting demand. In Sect. 3.1, we considered a case of 100 inbound flights of a single inbound bank all connecting to 100 flights within a single outbound bank, resulting in 10,000 hits. If we break down 100 inbound and outbound flights (see Fig. 34) into 10 daily banks, the resulting pattern will offer 10 banks, with each serving 10 inbound and 10 outbound flights. There are still 100 inbound and 100 outbound movements, but they are spread across 10 waves rather than within a single bank.

How many connections can we obtain in this structure (assuming BOF $= 1$)? Within each bank, the number of feasible connections is hits $=$ inb $\times$ outb or

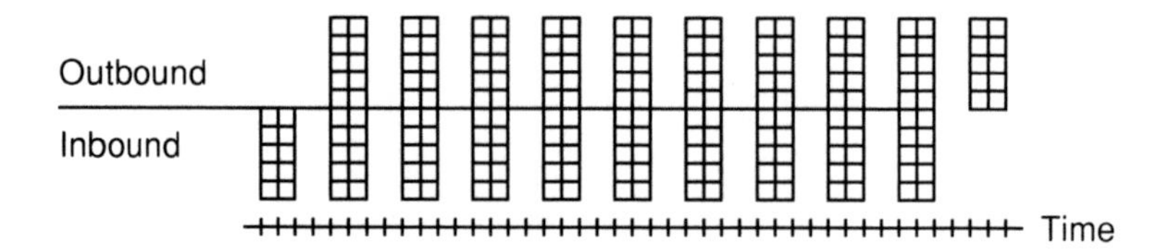

Fig. 34 One hundred inbound and 100 outbound flights distributed over 10 banks

$10 \times 10 = 100$. As we have 10 such banks, the total number of connections follows as 10×100 or 1,000. Comparing the number of feasible connections in the single-bank scenario (10,000) with the corresponding figure of the 10-bank pattern (1,000), the number of connections is reduced by a factor of 10—which is the number of banks we used. We obtain

$$\text{hits} = \frac{\text{inb} * \text{out} * \text{BOF}}{\text{banks}} \tag{7}$$

where banks is the number of banks or waves at the respective airport, and BOF is the bank overlap factor.

This formula may oversimplify the results in those cases where the various banks or waves serve a significantly different number of flights or a different number of directional flows. In cases where there are substantial differences in traffic structure between various banks over the day, formula (7) must be applied (without the denominator) to each individual bank, and the resulting hit scores per bank must be totaled.

Most of the large US hubs only serve two-directional flows—East-West/West-East. This improves connectivity, reducing the theoretical maximum of inb $\times$ out to a lesser extent than a 3, 4, or omnidirectional case (see Sect. 3.5). In contrast, the typical hubs in Europe are omnidirectional, significantly pushing down feasible connectivity. On the other hand, the still waved hubs (CO@IAH or NW@DTW) in the United States serve 10 or more waves, which run against maximum connectivity; while its European counterparts serve four (LH@FRA) or seven (AF@CDG) waves, offering some relief from the omnidirectionality of their traffic.

A sound bank structure must equally serve connecting passengers and local demand, including passengers departing or arriving at the respective airport as their origin or final destination. Local passengers typically demand departure and arrival times that are more evenly distributed over the time of day, and prefer to depart in the morning and return in the evening. From the connectivity point of view, serving a given transfer O&D twice or more during one bank is redundant for banks of relatively short duration. For that reason, a given origin rarely serves a given inbound bank more than once, and a given destination is rarely served multiple times during a given outbound bank. At hubs like FRA, which have few banks, this factor may compete with the need to serve prime destinations at high frequencies throughout the day.

The number of banks for optimum connectivity may be calculated by evaluating the average daily frequency of destinations. The objective is to match the number of banks with the number of daily frequencies per destination. Let us assume that 10 flights IAD-MSP simultaneously arrive at MSP, and 60 min later 10 flights simultaneously depart to SEA. This would not cause 10 × 10 hits, as 10 parallel inbound flights are equivalent to just one inbound flight with 10 times the capacity of the original flights. Because the same logic applies to the outbound side, 10 inbound and 10 outbound flights would still count as just one hit, not 100. This rule is called the "minimum hit criterion."

If we assume that MSP serves 10 waves—with each wave serving one IAD inbound flight and one SEA outbound flight, and the 10 waves are sufficiently distinct from each other—then we would count 10 IAH-MSP-SEA hits during the day. Each of the 10 waves offers one inbound IAH flight and one outbound SEA flight.

For a given number of frequencies F for a particular destination d and a given number of banks b, the maximum number of competitive hits may be calculated as follows:

$$\text{hits}_d = (\min(F_d, b))^2 \tag{8}$$

By adding up this score for all destinations d and the respective daily frequencies, we obtain the maximum number of hits for a given airport, destination/ frequency-portfolio, and the number of banks:

$$\text{hits} = \sum_{d=1,\dots,n} (\min(F_d, b))^2 \tag{9}$$

The higher the score, the more hits are competitive. By inserting different values of b, we can determine the number of banks for optimum connectivity for the underlying set of destinations and frequencies.

Figure 35 applies this formula to a few selected airline hubs, examining various scores of b (number of banks or waves). Take the NW at DTW case as an example. Apparently, the number of banks for optimum connectivity would be four, as compared to nine in reality. Contrary to popular belief, the number of banks or waves is not driving connectivity. It does so up to a maximum value, and any number of waves higher than that is counterproductive in terms of connectivity, provided the minimum hit criterion is applied. As seen in Fig. 35, most hubs show an actual number of banks much higher than needed if a maximum number of hits is the only objective. The reasons will vary from case to case:

- The number of optimum waves must incorporate other pivotal factors, such as operational feasibility, airport capacity, or slot constraints.
- With a higher number of banks, the corresponding connecting speed can be accelerated. If the competition is for fast connections, more and shorter banks might be advisable, even at the expense of the total number of hits. The spiked wave design of NW at its DFW hub suggests that this factor played a decisive role in the design of its nine-bank design, while four banks would have been enough to optimize the number of hits.

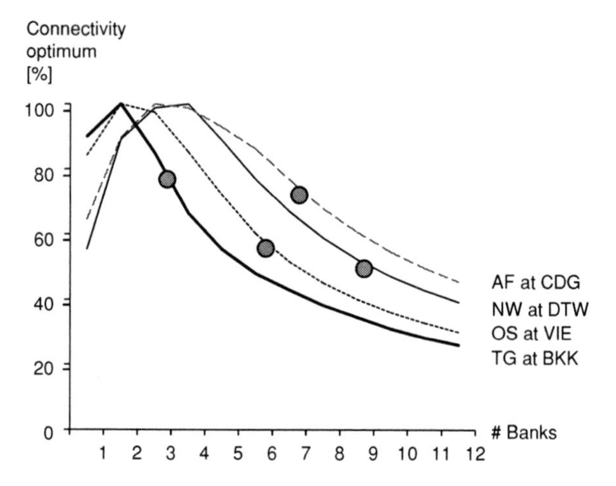

Fig. 35 Connectivity optimum and actual number (*circle* indicators) of banks for a selection of global hubs

- Passenger time-of-day preferences for local demand must be considered, as well as the convenience of the departure and arrival times of the underlying itineraries.
- With a lower number of banks, it may not be possible to achieve minimum levels of aircraft utilization. This factor is of particular importance for hubs serving many relatively short/medium-haul destinations.
- There is an underlying assumption inherent to formula (9): All banks are omnidirectional. However, if half of the banks serve one direction and the other half serve other destinations, the number of ideal banks must be multiplied by the number of relevant directions. This factor likely played a minor role in the examples given below, where nearly all banks serve the entire portfolio of directions (with the exception of a NW bank at DTW from 09:00 to 10:30, or an OS bank at VIE from 09:00 to 11:00, with both banks only serving one direction).

Given this interdependence between the optimum number of banks and connectivity, seasonal or other changes of frequencies in the flight program should cause respective adaptations in the wave or bank structure. If daily frequencies are significantly reduced for an extended time period in response to a cyclical downturn or a crisis, while maintaining an overly high number of banks, productivity is likely to suffer.

3.5 Connectivity Driver #4: Directionality

Few hubs serve all 360° of traffic. Most hubs specialize in serving a few selected directions. Fig. 36 shows the directionality patterns of key US, European, and Asian hubs. These graphs illustrate how a mega-hub like ORD is highly directional East–West, while DL's hub in ATL serves three dominating directions (South to Florida, North to the East Coast, and West to the rest of the US). CDG serves as an example of a typically omnidirectional European hub.

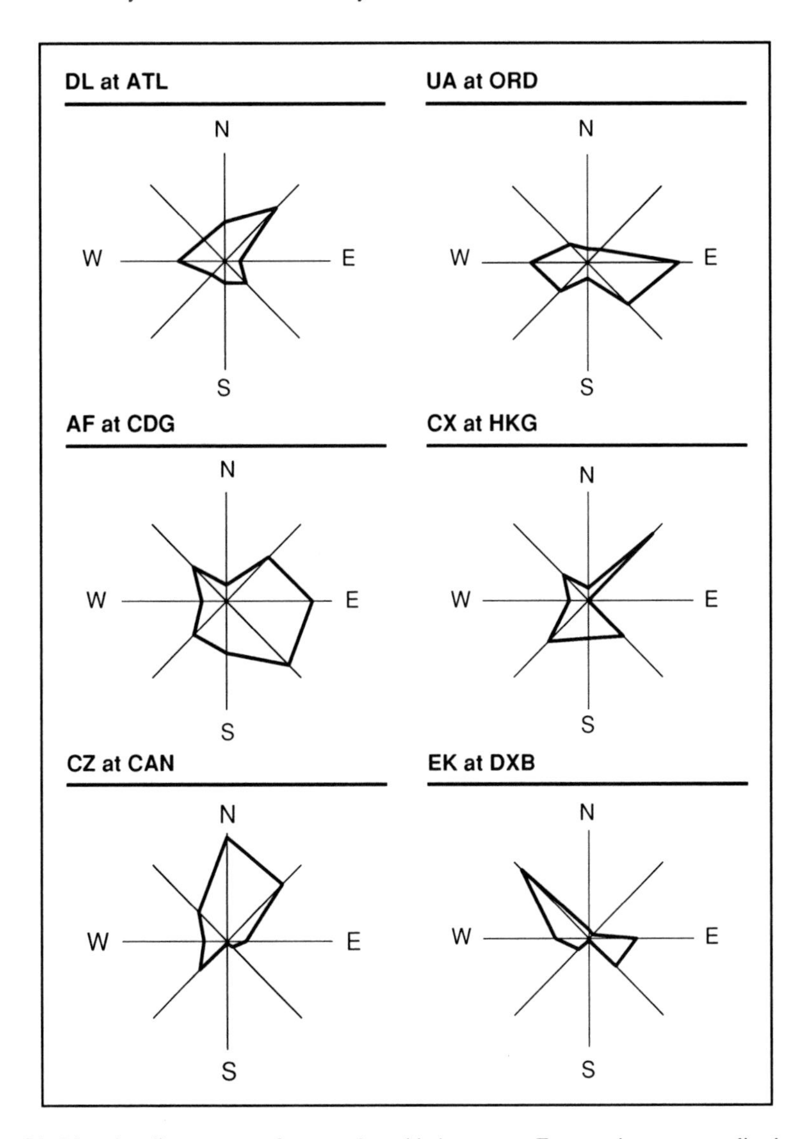

Fig. 36 Directionality patterns of some selected hub systems. Frequencies are normalized to the radius of the respective graph for each hub

One of NW's inbound waves at MSP is arriving between 10:00 and 10:45 local time, corresponding to the outbound wave from 11:00 through 12:00 (see Fig. 37). The inbound wave is composed of flights mainly from the East, whereas the outbound wave is built up of flights mainly departing to the West. Having a highly directional inbound bank and a highly directional outbound bank with a fitting counter-direction is the most basic type of wave design.

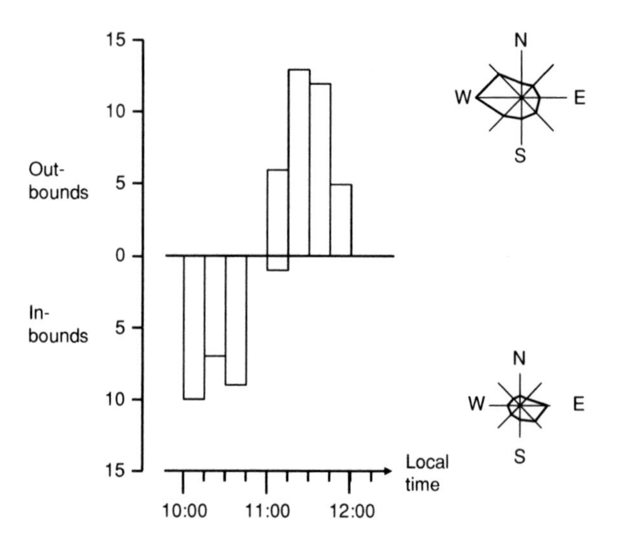

Fig. 37 Wave pattern and directionalities of NW at MSP. The inbound bank is composed of flights mainly from the East, while the outbound bank serves flights mainly to the West

If a particular wave has to serve more than two complementary directions, the increasing number of directions starts to negatively affect connectivity. Let us assume that the inbound flights to each wave are composed of five westerly and five easterly inbounds and that the same applies to the outbound flights. All inbounds can no longer connect with all outbounds within a single wave. Five inbounds from the West can connect with five outbounds to the East, and five inbounds from the East can connect with five outbounds to the West. In this example, the number of feasible connections per wave is reduced to $hits = 2 \times 5 \times 5$, or 50, a further reduction by two [again, assuming the BOF (bank overlap factor) to be = 1].

The directionality issue becomes slightly more complex when we expand the "two directions" example (East, West) into three or more directional segments (see Fig. 38). Let us assume an inbound wave with three destinations, one coming straight from the North (0°), another from the Southeast (120°), and one from the Southwest (225°). Each of these directional segments is served by just one flight, and we find the same pattern in the corresponding outbound wave; each of the three destinations can connect with the other two destinations within reasonable detour ranges. Then, any inbound flight can only connect with two out of three outbound flights, forming a connectivity reduction factor of 2/3. In general terms, the directional reduction factor is defined as:

$$r = \frac{facc}{fav} \tag{10}$$

where r directional reduction factor, $facc$ number of accessible directional outbound segments, fav number of all available directional outbound segments.

To give an example: An inbound from the South (Florida) to ATL can build connections with two (Northeast and West) of the dominant outbound directions (South, Northeast, and West). Connecting back to the South is not feasible. Hence, the directional reduction factor for DL at ATL is about 2/3 (assuming that West

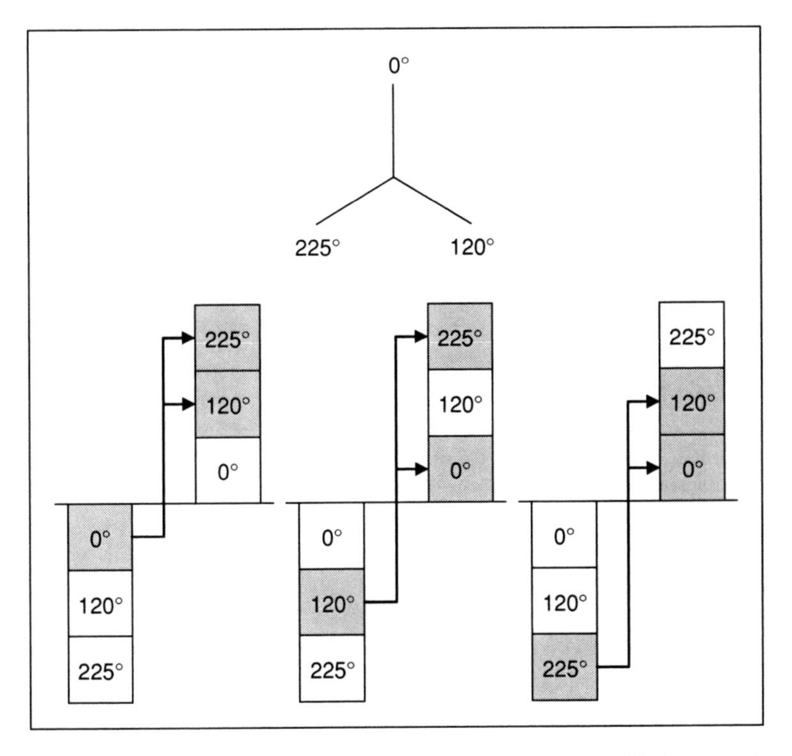

Fig. 38 The impact of directionality on connectivity. Consider a central hub, connecting destinations at 0°, 120°, and 225° (above). Let us further assume that each of the three destinations can connect with each of the other two destinations within reasonable detour ranges. Then, any inbound flight can only connect with two out of three outbound flights, forming a connectivity reduction factor of 2/3

inbounds can connect to South and Northeast outbounds, and Northeast inbounds can connect to South and West outbounds).

Integrating this aspect into the "overall" formula gives us

$$\text{hits} = \frac{\text{inb} * \text{out} * \text{BOF} * r}{\text{banks}} \tag{11}$$

Let us apply this formula to the earlier example of inb = 100, out = 100, banks = 10, bank overlap factor (BOF) = 1 example (see Sect. 3.4). Assuming that directional traffic flows East and West, we can build directional flows East–West and West–East ($r = 1/2$, as one inbound directional cluster can build connections with one out of two outbound directional clusters). Then, we can calculate the resulting total number of hits as:

$$\text{hits} = 100 \times 100 \times \frac{1}{1} \times \frac{1}{10} \times \frac{1}{2}$$

$$\text{hits} = 500 \tag{12}$$

In Sect. 3.4, the number of hits for the case of 100 inbounds and 100 outbounds, distributed over 10 banks, and no consideration of directionality, was assessed as 1,000. For the same parameters, but considering directionality, the number of hits goes down by half.

3.6 Connectivity Driver #5: Rotational Patterns

The number, timing, and directionality of banks at a given airport are greatly influenced by the opportunities offered or constraints imposed by typical stage lengths and the resulting aircraft rotational efficiencies. As a result, the bank design has a great impact on aircraft productivity. High aircraft utilization requires the tight fit of an aircraft's rotational pattern with the opening hours of its home-base airport and as short as possible turnaround times. Thus, a portfolio of destinations with flight stage lengths permitting a high number of outbound and homebound flights per day positively affects aircraft utilization. Moreover, stage lengths must permit rotational patterns so that all aircraft arrive at a particular hub at exactly the optimum arrival time of an inbound bank. If the timing of a bank and the required flight times to or from a specific destination do not match, aircraft idle times and reduced aircraft utilization are the unavoidable consequences. To give a positive example (see Fig. 39), assume a single hub network with two spoke destinations, A and B, being away from the base either 60 min (destination A) or 120 min (destination B) block time. The shorter rotation requires a TAT of 30 min, and the longer rotation requires a TAT of 60 min. Airport opening hours are from 06:00 through 23:00 local time. The resulting rotational patterns would lead to two major banks (11:30–12:00, 17:30–18:00) and three additional minor banks (08:30–09:00, 14:30–15:00, 20:30–21:00).

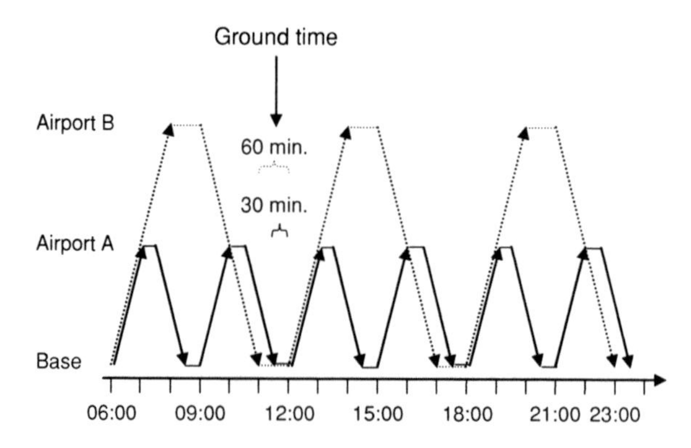

Fig. 39 Stage length, aircraft utilization, and bank design. Schematic example of the relationship between prevailing stage lengths, aircraft utilization, and efficient bank timing for a network of a base hub, destination A (60 min block time, TAT 30 min), destination B (120 min block time, 60 min TAT), and base hub opening hours from 06:00 through 23:00 (local time)

Figure 40 shows the same case as in Fig. 39, but assumes a 50-min stage length rather than 60 min for flights from the base to destination A. As a result, flights of the rotation base-A-base must stay 50 min on the ground each time in base rather than only 30 to build optimal connectivity with the base-B-base flights. Over the day, this adds up to 100 min of unnecessary idle time for this individual rotation. If we assume a fleet of 100 aircraft in a network, and all suffer from a mere mismatch of 10 min, idle times add up to a capacity equivalent of one aircraft. This example makes clear the enormous impact that the bank design may have on aircraft productivity.

In reality, the actual portfolio of various stage lengths rarely matches a particular bank design in an ideal way. Compromises must be made in terms of connectivity or productivity. Note that idle time in rotational plans not only affects aircraft productivity, but also impacts crew and other asset productivity.

Figure 41 shows the distribution of stage lengths for a major European carrier at its prime hub. In an ideal case, such a distribution shows a distinct temporal distance between the various spikes. Let us reconsider the example given in Fig. 39, with one circle of 120-min stage length and a 60-min TAT, and another circle of 60-min stage length along with a 30-min TAT. Given these proportions of circle time and TATs, stage lengths of 60 min and 120 min permit building rotations of optimum aircraft productivity and optimum connectivity. A different TAT or stage length would compromise productivity, connectivity, or both. Apparently, such ideal proportions are rarely the case in real life. Let us assume a network planner at the airline represented in Fig. 41 decides to build a rotation based on the fact that most flights take about 120 minutes (see dotted line in Fig. 42). For the sake of simplicity, let us further assume that a TAT of 60 minutes applies to all flights. How do we fit in the flights whose stage length is shorter than 120 min (see solid line in Fig. 42)? These flights must stay on the ground longer than required by TAT to optimally connect with the dominating group of flights of 120-min stage length.

The shape of the stage length portfolio greatly enables, or limits, the feasible levels of aircraft utilization. Many airlines therefore blend two or more bank systems into one by delaying the starting point or termination of one bank system

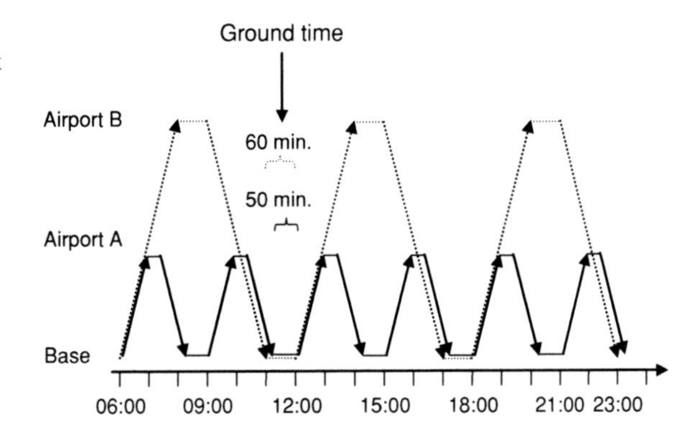

Fig. 40 Same scenario as in Fig. 39, but assuming a block time of 50 min, rather than 60 min, for base-A flights (*solid line*)

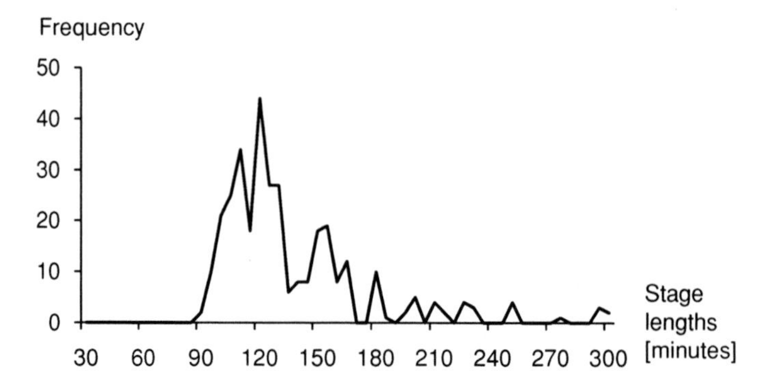

Fig. 41 Stage length profile for the hub of a major European airline. The frequency of flights over stage lengths is shown

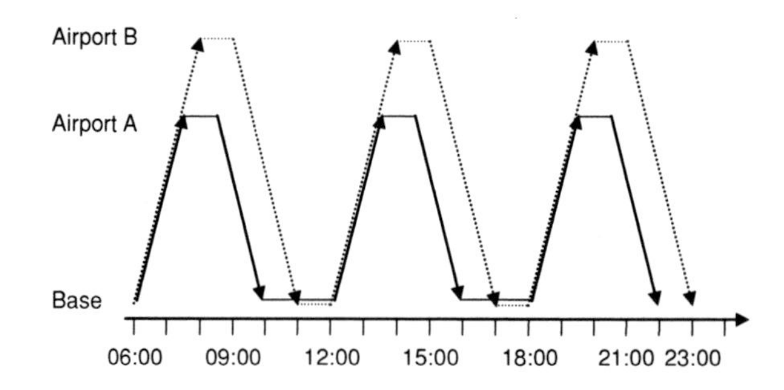

Fig. 42 The longer stage length (*dotted line*) dictates the idle times of slower rotations (*solid line*)

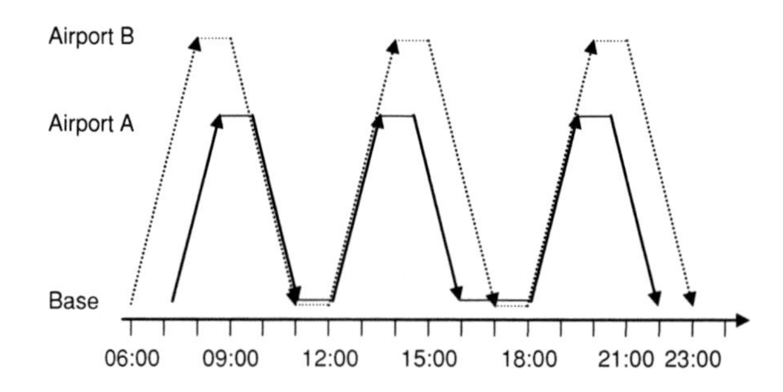

Fig. 43 Blending two rotational patterns into one. The stage length of the rotation behind the *solid line* does not permit an efficient combination with the stage length forming the *dotted line*. In this case, the starting time of the solid rotation was delayed to permit connectivity at the expense of idle time

against the others. This does not eliminate the productivity disadvantage but may contribute to connectivity, as well as a more even distribution of connections over all times of day (see Fig. 43).

In reality, virtually no waved system is perfect in the sense of optimum aircraft rotations and aircraft utilization. Some airlines respond by adding idle times into

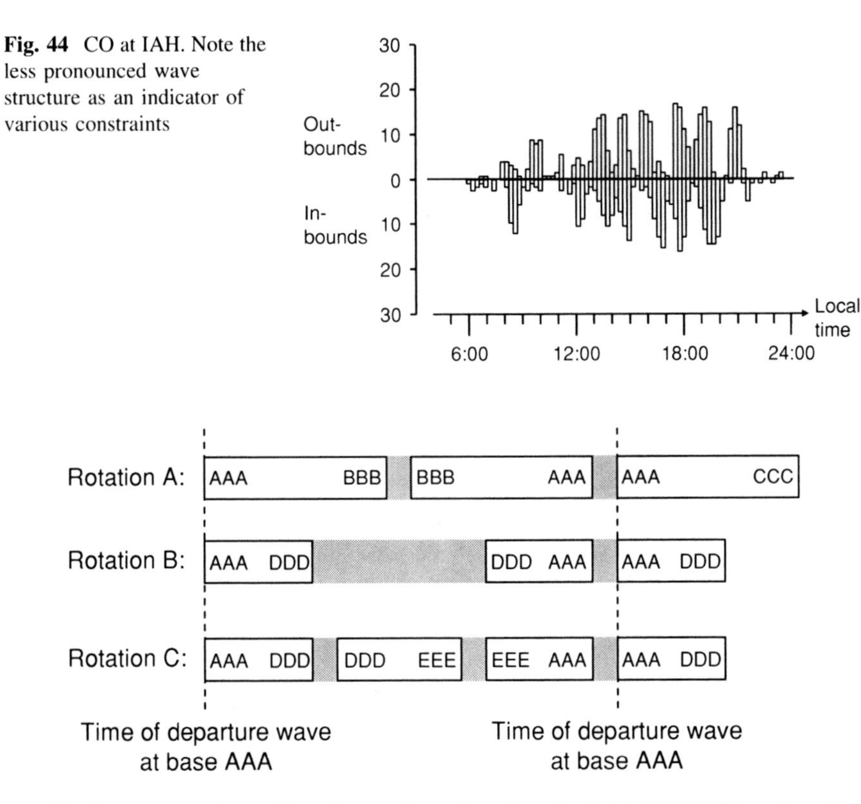

Fig. 44 CO at IAH. Note the less pronounced wave structure as an indicator of various constraints

Fig. 45 Triangle rotations. Triangle rotations (Rotation C: AAA-DDD-EEE, rather than AAA-DDD as in rotation B) offer more opportunities to schedule efficient arrival times at the waves in hub AAA. The *white bars* denote flights; the *gray bars* denote ground time

their schedule to accommodate timing requirements of waved hubs; others accept less clearly shaped, "flattened out," or even overlapping bank structures. Most often, however, a sharply waved pattern of activity at a hub is a disadvantage in terms of feasible aircraft utilization. Generally, a clearly waved hub structure and high asset utilization are not complementary. Figure 44 uses CO at IAH as an example of a compromise between a spiked wave structure and operational or other requirements.

For many years, airlines have attempted to solve the problem of inefficient aircraft rotations in waved hub environments by replacing ping-pong (out-and-back) rotations (see Sect. 2.2 in Chap. 2) with triangle or complex rotations before arriving at a waved hub (see Fig. 45). Sequencing flights across various airports provides greater opportunities to find a leg sequence that offers ideal arrival times at a waved hub.

By disrupting local flights, however, the sequencing of rotations across various hubs can easily cause widespread delays. If the first leg of a tightly packed rotation covering several airports is delayed by 15 min, and all airports along the planned rotation are hubs with spiked waves, this single delayed flight will affect the

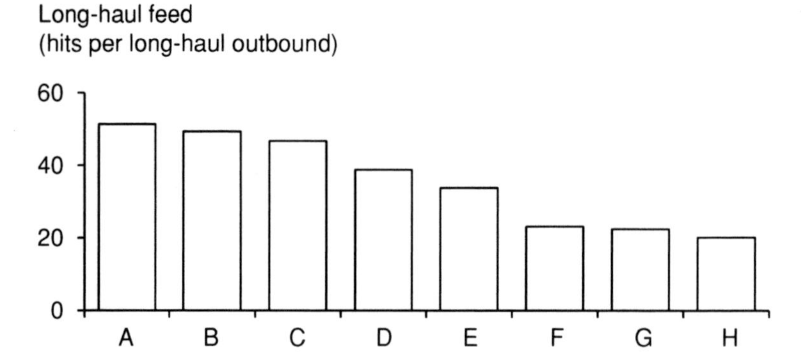

Fig. 46 Feeder leverage for long-haul operations at selected European hubs. A high or low score is the result of the strategy applied, as well as performance

dependent connections at all subsequent hubs, again causing and spreading delays. The result is an exponential growth of delayed flights across the network within minutes. One way to anticipate such disruptions is to build sufficient time buffers into all rotations. However, this would be costly in terms of asset utilization. Another option is for airlines to contain the local impact of delays by planning as many ping-pong rotations as possible. As part of its "Operation Clockwork" in 2004/2005, DL increased the share of ping-pong rotations at its ATL hub to over 92% (Petroccione 2007).

Long-haul flights are fairly constrained in terms of arrival and departure times. Time-of-day profiles at both the origin and destination must be considered, as well as efficient TATs at the outstation and airport curfew constraints. At the same time, long-haul flights typically carry a large number of connecting passengers, and thus require more feed and de-feed connectivity than short/medium-haul flights. At hubs with a high number of long-haul flights, those flights determine the number and timing of banks. In such cases, banks are typically designed to optimally accommodate long-haul flights.

To demonstrate the interdependence of long-haul flights and connectivity, Fig. 46 compares the number of feeder flights observed for long-haul outbounds to a sample of selected European long-haul hubs.

3.7 Connectivity Driver #6: Airport Infrastructure

Various network strategies and structures depend on the specific airport infrastructure and the efficiency of related commercial or operational procedures. This applies equally to terminal, apron, taxiway, runway, and airspace design. Key questions must be addressed: How do we assess overall efficiency? How long does it take to approach, land, and taxi until "on blocks"? How long does it take from "off blocks," push back, taxiing, take-off, and climbing up to be back

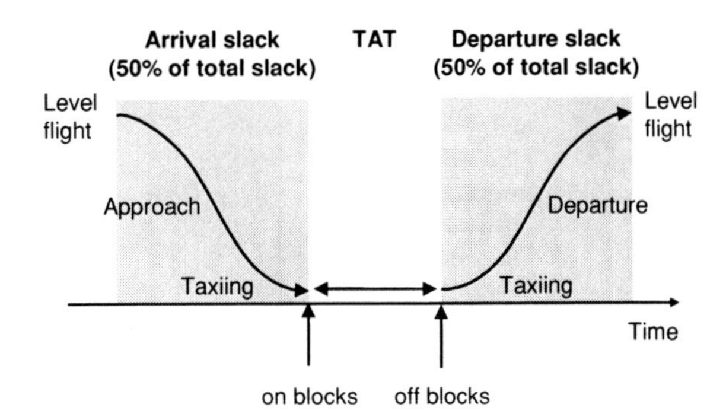

Fig. 47 Airport slack measures the airport-specific total time between the beginning of approach and "on blocks" plus the total time between "off blocks" and completion of climbing

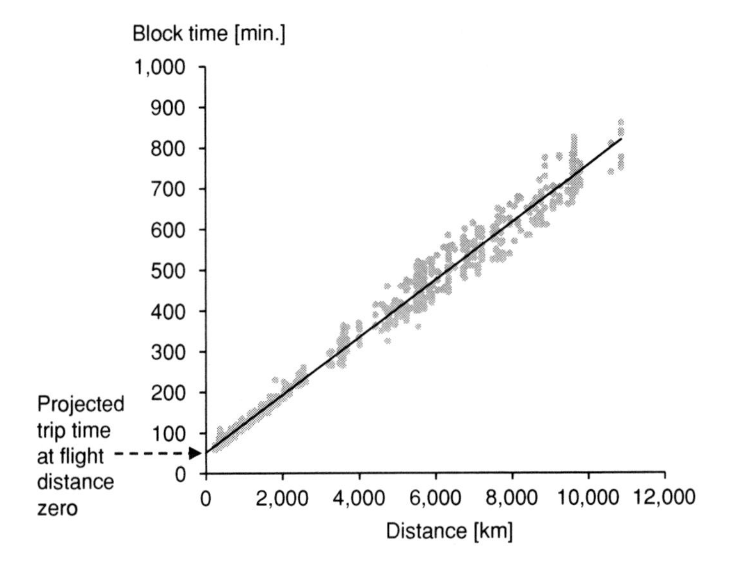

Fig. 48 Airport slack for LHR. For each individual flight movement (*dot*), the flight distance is plotted against the respective block time. The projection of the linear regression of these values to a flight distance of zero indicates the typical time required for a complete approach and take-off cycle at this airport

en route (see Fig. 47)? The time in between, TAT, usually is well understood and optimized. What about the other time periods mentioned, however? How long do they take? Can they be determined and optimized? How do they compare among airports?

The most straightforward approach to assessing total airport slack is to compare the block times and distances of all inbound and outbound flights at a given airport. The regression line "block time versus distance" of these data do exhibit a significant correlation (see Fig. 48). If we extrapolate this regression to a flight distance of zero, the corresponding block time will be larger than zero, however.

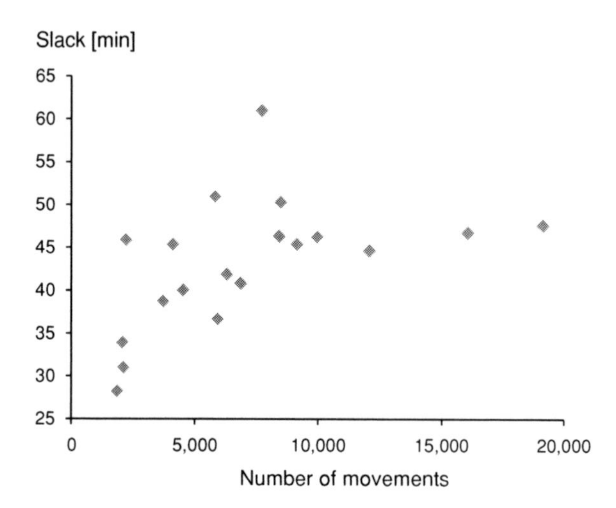

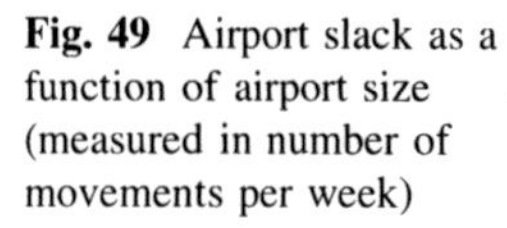
Fig. 49 Airport slack as a function of airport size (measured in number of movements per week)

The block time at zero distance is the time that a flight planner has to consider as the time needed for a complete approach and the departure process at the given airport, net of TAT. This time period is called "airport slack."

In Fig. 49, we plot the overall size of hubs with the corresponding slack. While an overall correlation appears to be weak, the data do confirm that:

- Being small is a precondition for an airport being fast.
- Growing in size lowers the chances of an airport being fast.

MCT is a measure of terminal process efficiency, and TAT is a measure of ramp process efficiency. Slack covers the entire spectrum of airport-specific and typical times needed for holding patterns, approach, and climbing processes, net of TAT.

3.8 Connectivity Driver #7: Random Connectivity

Strictly speaking, randomness is an important driver of connectivity, as an underlying number of connections can be expected as the result of a large or small number of flights randomly distributed over time [formula (4)]. Some of the largest hubs worldwide bet on randomness as the prime driver of connectivity. Randomness rarely leads to the highest scores of connectivity, however. For large hubs, connectivity may become a by-product of hub structures that emphasize productivity. For this reason, random-based hub structures will be discussed in Chap. 4. It is important to understand that pure random structures do not permit giving priority to any types of hits. Therefore, many network planners first plan selected flights, such as long-haul or high-value O&Ds and destinations (see "pillar flights," Sect. 4.7 in Chap. 4), and then "pour in all other destinations randomly, like filling water into a lake after you have defined some islands in the lake," as one has put it.

3.9 Connectivity Driver #8: Minimum Connecting Time (MCT)

A competitively short MCT is a vital precondition to building competitively fast connections. However, there is no universal standard of how fast an MCT must be to permit competitive hits. For many transfer O&Ds in Asia, nearly all potential transfer points allow fairly slow MCTs. In Europe and particularly in the United States, competitive levels of connectivity require aggressive MCTs. It is a widespread misconception that adding "only" 5 min to an applicable MCT to impose additional regulatory constraints would have a marginal effect. If two hubs compete as transfer points for the same transfer O&Ds, why would passengers opt for the slower connection? Also, the longer the MCT, the lower the number of random hits. As more hub structures are built upon random hits in exchange for higher asset productivity (see Chap. 4), slower MCTs will counter connectivity and productivity.

3.10 Connectivity Driver #9: Internal Structure of Banks

The allocation of flights within an inbound or outbound bank is usually driven by operational efficiency considerations, directionality, the respective value of individual flights, or a combination of these factors. In Sect. 3.5, we reviewed the importance of directionality to connectivity. Given the importance of directionality in building high-performance transfer systems, the allocation of flights for a particular direction within a bank can play a significant role. Apart from directionality, the deployment of long-haul flights within an inbound or outbound bank also has a significant impact on connectivity. The positioning within a bank of flights representing high or low average yields may contribute to the competitiveness of the bank system.

3.11 Timing of Long-Haul and Short/Medium-Haul Flights in a Bank System

To optimize operational robustness, increase passenger convenience, and keep connectivity at competitive levels, some important rules on structuring waves are required.

Consider LH's evening bank (see Fig. 50) at its FRA hub, ranging from 19:00 (inbound) through 23:30 (outbound). Note that long-haul flights (gray) arrive early

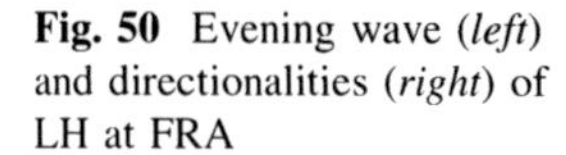

Fig. 50 Evening wave (*left*) and directionalities (*right*) of LH at FRA

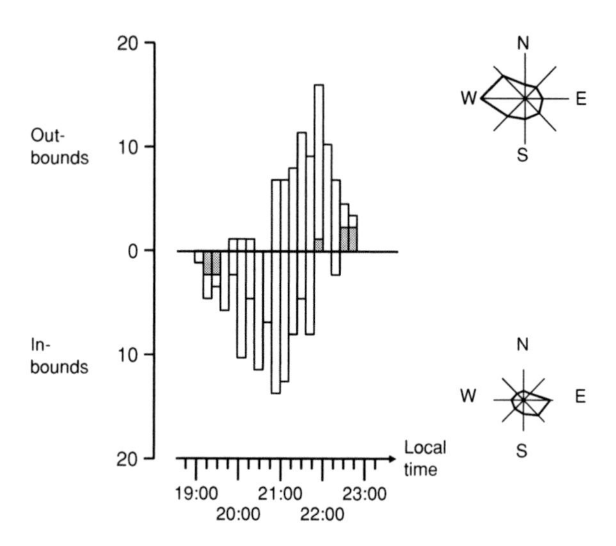

in this bank and depart late in the corresponding outbound bank. The rationale for this timing is twofold:

Passengers arriving on long-haul flights require significantly more time to get to the departure gate of their connecting flight, as customs, baggage handling, and other procedures absorb time that does not equally apply to domestic or short/medium-haul connections. Long-haul flights typically are nondomestic, so generally undergo time-consuming security and customs procedures. As a result, the MCT of international connections is usually much longer than that of domestic connections. To provide the necessary time span within a time-limited, single inbound/outbound wave combination, long hauls must arrive early and depart late.

From an operational point of view, long-haul aircraft must arrive early in a bank and depart late. As a rule of thumb, all aircraft arriving during an inbound bank should be able to depart again during the corresponding outbound bank. Because TATs for long-haul aircraft are much longer than TATs for smaller single-aisle aircraft, long-haul aircraft can only arrive and depart within one wave if arriving early and departing late.

Conversely, if long-haul inbounds are to arrive early in an inbound bank and depart late from the outbound bank, the proportions for short/medium-haul flights are the opposite in order to build faster connections. A flight arriving in the second half of the inbound bank and connecting to a flight during the first half of the outbound bank is faster than any other configuration. Competition on minor time advantages is more pronounced for short/medium-haul connections than for longer-haul connections.

At hubs combining a high number of long-haul flights with a high number of short/medium haul-feeder and de-feeder flights, network planners usually try to plan long-haul connections in an asymmetric way. Long haul to short/medium haul is scheduled using faster connecting times, close to MCT. Short/medium haul to long haul, however, is given significantly more time to connect. The reason for this asymmetric proportion is simple: If a long-haul inbound passenger misses a tight connection to a short/medium-haul outbound flight, the probability is high that the next flight to the missed destination is only a short time away. Conversely, a missed connection to an outbound long-haul flight is not likely to soon find an alternative flight. Thus, more buffer time must be given to long-haul outbound connections than to long-haul inbound connections.

Short-haul to short-haul connections are either fast and risky due to mild delays or prohibited by design. Many network planners place flights in close proximity in order to miss MCT. This is due to their typically low yields and relatively high environmental impact.

3.11.1 Structuring Banks and Hubs by the Value of Connections

Hits may differ along operating costs, yield level, price elasticity, demand volume, competitive intensity, growth potential, volatility, and available capacity. At the end of the day, some hits offer more value than others. This fact has a deep impact on the portfolio of destinations served, the corresponding frequencies and capacities deployed, and the resulting network, hub, and bank structures. For instance, AA moved away from a strategy connecting "everything with everything" at DFW and ORD in 2002 to focus on more valuable East–West flows. A structural consequence was that AA gave up its sharply waved hub structure at its hubs in DFW and ORD and reverted to flatter structures. Some airlines prioritize high-value hits within bank structures over hits representing lower value. While such a policy is likely to attract more valuable traffic, it also introduces greater complexity and more complexity costs.

There is no general rule of thumb on how to respect differing hit values when designing network, hub, or bank structures. However, all the methodologies and formulas offered throughout this book still apply: The key is to properly phrase the strategic question. For example, imagine the ideal bank structure for a portfolio of high-value and low-value flights and traffic flows. In the framework of an undifferentiated portfolio of flights, the question would be: "What bank structure provides the maximum number of hits?" In the context of a differentiated portfolio of traffic, however, the focus is on finding the optimum structure for high-value flights and traffic flows, and then adding flights of lower priority. The methodology of designing bank structures in both instances is the same.

One caveat: The more structure is added to a hub and bank, the more vulnerable the structure is likely to become to operational delays. Accommodating different hit values when designing bank structures is likely to add operational instability. Planners must build schedules that offer not only high commercial appeal, but also operational robustness. That is why most airlines are avoiding over-sophisticated bank designs and moving toward more random and flattened structures. This trend may offer particular advantages for large hubs and those experiencing strong growth.

3.11.2 Connectivity and Operational Robustness: A Contradiction?

Any hub structure, be it banking or the internal structure of banks, can constrain operational robustness. Note that operational robustness differs from operational efficiency. The more dependent a detailed time structure is on a schedule structure, the more vulnerable the structure becomes to delays. This holds true for the internal structure of banks as well. The more rules that apply—such as "long-haul arrives first inbound" and "pillar destinations have strict priority as long-haul de-feeders"—the more likely that delays will destroy the rationale of the overall structure.

Chapter 4
Designing Asset-Productive Networks

Abstract Optimum asset utilization is vital for airlines and airports alike. The productivity of the assets deployed is at least as important as commercial objectives like connectivity. Certain schedule characteristics are more likely to facilitate high utilization of resources, while others are more likely to reduce it. Regardless of the various hub structure tactics—peaked, de-peaked, waved, rolling, or random—infrastructural constraints can significantly impact efficient asset utilization. One of the most severe infrastructural constraints is the opening hours of airports. Efforts to build time buffers into airline timetables to increase operational stability are likely to impose additional constraints to efficient asset utilization.

Connectivity affects the revenue side of the airline business model equation, while productivity is clearly a cost issue. In times of expanding markets, airlines emphasize connectivity; but during times of traffic congestion, productivity becomes a more prominent issue. As a general trend, however, the pure-play connectivity schedules increasingly were replaced by structures that take a more balanced approach between connectivity and productivity. The larger a hub system grows, the less important connectivity becomes (see Sect. 4.2), and the more important asset utilization becomes as a profitability lever. This explains why more airlines seek network and hub structures that support asset productivity and maintain connectivity at appropriate levels.

4.1 Aircraft Utilization Revisited: Why Asset Productivity is Vital for Profitability

4.1.1 How to Measure Aircraft Utilization

Aircraft utilization is measured as average daily utilization of all aircraft within a given fleet. It is determined by two factors: the number of aircraft needed to operate a

P. Goedeking, *Networks in Aviation*, DOI: 10.1007/978-3-642-13764-8_4,

given schedule, including operational reserves, and the number of flight hours flown per day (measured as block hours, abbreviated as Bh). Airlines first calculate the number of aircraft needed from their internal rotation plans. This direct measure of fleet requirement is a result of the schedule design and the emphasis given to aircraft productivity when designing the schedule. The real fleet size, however, exceeds this direct measure, as a certain number of standby aircraft must be available in case of an operational disruption. The number of such "ops reserve" aircraft depends on the respective airline's fleet structure, on-time performance history, service quality, and other factors. When airline fleet utilization data are made publicly available, they include operational reserve as the "nominator" (the number of aircraft in the fleet), they are grouped by the level of "narrow-body" or "wide-body" aircraft, or they are infrequently published with significant delay. A simple way to assess the minimum number of aircraft needed to operate a given schedule (the program fleet) is by counting the number of aircraft in a given fleet that are airborne at any specific time over an entire week. The maximum of that score is the minimum aircraft needed to operate this schedule, not including any operational reserve.

4.1.2 Bank Structures Significantly Impact Aircraft Utilization

Most connectivity drivers discussed in Chap. 3 impact the utilization of aircraft, crew, terminal, or other assets. The bank design shown in Fig. 21, for instance, will surely lead to extended ground times for all aircraft, crew, and gates involved. Aircraft turnaround times are generally faster than passenger MCTs, except if significant maintenance work is needed. If the bank design is tuned toward optimum connectivity—to avoid overlap between an inbound and corresponding outbound bank, for instance—such connectivity will come at the expense of sound asset utilization. If the individual banks are separated by significant pauses, aircraft most likely will have to wait at the outstation before the next available inbound wave at the destination hub can be met, again curtailing asset utilization. As a general rule of thumb, connectivity is optimized at the expense of asset utilization, and asset utilization is optimized at the expense of connectivity.

4.1.3 Operational Standardization Impacts Aircraft Utilization

LCCs, at least in their pure-play format, aim to minimize production costs as much as possible, translate such production cost advantages into competitive price advantages, and attract budget-conscious demand. The key to minimum production costs is simplicity and standardization. Pure-play LCCs operate one standard type of aircraft in all the same configurations and disregard connectivity or any service beyond the essential minimum. Strict standardization is a precondition for zero-fault operational procedures, which are used as the best replacement for otherwise unavoidable time buffers. However, this is very different from what some network carriers refer to when improving the operational robustness of their networked schedules. For them, operational robustness translates into time buffers built into the schedules to compensate for likely delays and to reduce productivity. Before studying the various levers employed by network carriers to stabilize the operational delivery of complex schedules, we will analyze the standardization and productivity optimization mechanisms that are built into simplicity-driven LCC networks.

4.1.4 Infrastructure Availability Drives Aircraft Utilization Up or Down

The most effective driver of airline productivity is aircraft utilization, as an aircraft sitting on the ground cannot make money. Consequently, LCCs select bases, destinations, and aircraft types, and define boarding, disembarking, and turnaround procedures all around the overarching objective of highest daily utilization of aircraft. Assuming that TATs are identical at the outstation and the base, and for the outbound and return flights, we can write:

$$t_e = (\mathrm{SL}_i + \mathrm{SL}_o + \mathrm{TAT}_b + \mathrm{TAT}_o) \tag{13}$$

where t_e is total excursion time, SL_o stage length of given outbound flight (minutes), SL_i stage length of given inbound flight (minutes), TAT_b TAT at base, TAT_o TAT at outstation.

For practical purposes, formula (13) can be simplified as

$$t_e = 2 \times \mathrm{SL} + 2 \times \mathrm{TAT} \tag{14}$$

Assuming an airport with opening hours from 06:00 through 24:00 (=18 h, or 1,080 min), a TAT of 45 min, and an SL of 90 min, this would result in exactly four full cycles:

$$\frac{1{,}080}{2 \times 90 + 2 \times 45} = 4.0 \tag{15}$$

with a TAT of 50 min, or 5 min longer, only three cycles would be possible:

$$\frac{1{,}080}{2 \times 90 + 2 \times 50} = 3.86 \tag{16}$$

The remainder of 0.86 translates into an idle time of 240 min, which negatively impacts the desirable productivity objective. For an LCC, airport opening hours must be the maximum possible, but at some point in time will be given. TAT must be the minimum possible, but at some point in time will be defined as fixed. The only remaining variable is flight time, or SL. In the example given above, a flight time of 120 obviously yields optimum results. What other flight times achieve the same or similar levels of productivity? Considering that the ultimate objective is to minimize idle time, we can write (assuming all *j* flights show identical block time and that the same TAT applies to all turns at the base):

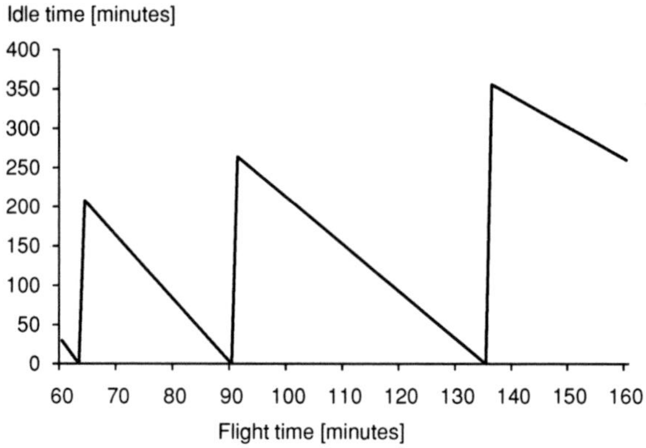

Fig. 51 Aircraft utilization and flight time. Productivity of an aircraft as the result of various flight times, assuming airport opening hours (06:00–24:00) and TATs (45 min) as constant (*x*-axis: flight time, *y*-axis: resulting idle time)

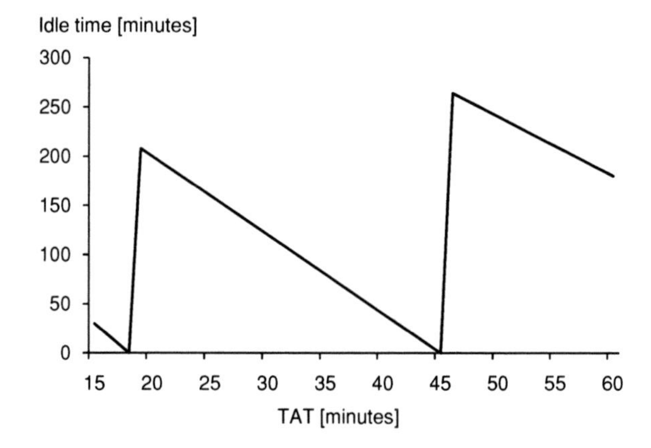

Fig. 52 Aircraft utilization and TAT. Productivity of an aircraft as the result of various TATs, assuming airport opening hours (06:00–24:00) and flight time (90 min) is given (*x*-axis: TAT, *y*-axis: resulting idle time)

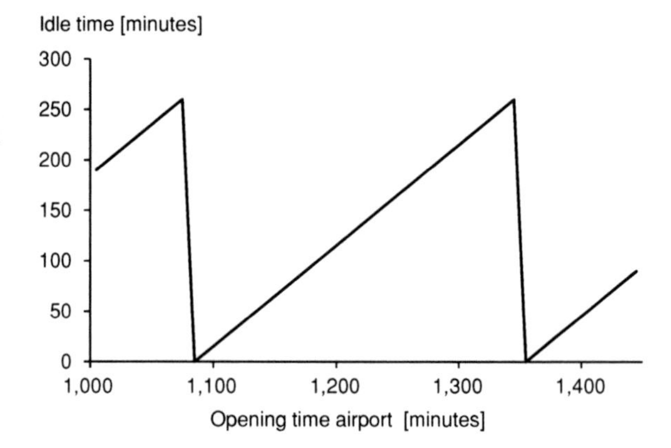

Fig. 53 Aircraft utilization and curfew. Productivity of an aircraft as the result of various airport opening times, with flight times (90 min) and TAT times (45 min) kept constant (*x*-axis: opening hours, *y*-axis: resulting idle time)

$$\text{minimize } (T \text{ modulo } 2 \times \text{SL}_j + 2 \times \text{TAT}) \tag{17}$$

where T is opening hours (in minutes), j flight index.

Applying this formula yields flight times (SL) of 63 min for five cycles, 90 min for four cycles, and 135 min for three cycles as the productivity optimum.

Figure 51 shows the idle times resulting from other flight times; Fig. 52 shows the idle times for various TAT values for a given flight time; and Fig. 53 demonstrates the productivity impact of seemingly minor reductions in airport opening hours.

Clearly, even seemingly minor changes of applicable constraints can have an immediate and costly impact on productivity.

4.1.5 How Bank Design, On-Time Performance, and Operational Robustness are Interdependent

Bank or wave designs, if planned too tightly to minimum values of MCT and ground times/TATs, are more vulnerable to delays than structures offering more buffers. Figure 21 gives an example of a bank design bound to incur delays, as compared to a more buffered design. The rectangular bank causes a sudden and steep increase in connections immediately after completion of MCT. That means that even a mild delay will destroy many connections by falling below MCT. A softer shape at the end of the inbound wave and at the onset of the outbound wave generates a more moderate increase in the number of connections. Providing more buffer helps to avoid delays, maintain minimum room to maneuver, and fix delays via circular swapping (see Sect. 2.2 in Chap. 2). Through circular swapping or other recovery mechanisms, a limited number of disruptions or delays can be repaired, but likely not for the majority of connections. Internal structures of banks, such as emphasizing certain destinations or types of destination (direction, value, volume) at particular times within an individual bank all make the bank design more vulnerable. Aside from the design of bank structures, three mechanisms help to increase the operational stability of schedules:

4.1.5.1 Buffering Flight Times

Network carriers are increasingly confronted with flight delays, and consumer organizations regularly publish delay statistics per airline and hub. Airlines initially responded by extending off-block time of critical flights. After adding buffers to "air time," they found that extended off-block flight time was mysteriously reabsorbed as flight time, still resulting in too tight ground times. Buffered flight time often led to a down-prioritization when approaching a congested airport and to less favorable taxiing routes or gate allocations. As a result, airlines started to extend time parameters on the ground rather than in the air, such as TATs of aircraft and MCTs for passengers.

4.1.5.2 Buffering Ground Times

As an alternative to building buffers into flight times, an increasing number of airlines are extending planned ground times. This way, the buffer does not need to become evident to any institution outside the airline itself, preventing buffers from being absorbed for other purposes than the intended one. In practical terms, minimum ground times, or TATs, for the various types of aircraft are given an extra 5 min or so, more likely at the hub rather than at a spoke or outstation. Buffered ground times appear to be more effective than buffered flight times of the same duration; and hidden buffers appear to be more effective than visible buffers.

4.1.5.3 Inserting Recovery Gaps

In the United States, some airlines build 15-min "recovery gaps" (Petroccione 2007) into their schedule to separate high-traffic outbound periods. These recovery gaps offer a ground-time buffer to recover from disrupted operations before the start of a new bank or before a period of intense or high-value operations. That is why such recovery gaps are found only on the outbound side of operations.

4.1.6 Flat Hub Structures: A Revolutionary Innovation or a Surrender to Complexity?

From the onset, LCCs have relied on simple, standardized, and operationally efficient network and hub structures. Beginning in the mid 1990s, network carriers started to experiment with borrowing some elements of the LCC hub structures to escape the mounting complexity costs of predominantly connectivity-driven structures. The prime objective was—and still is—to improve the productivity of their key assets, namely their crew, aircraft fleet, and ground assets. To achieve that objective, the spiked structures of waved hubs were key suspects in undermining productivity. Adding idle time to flight schedules led to congestion at gates, taxi-ways, runways, and in the air. As a result, "de-peaked" or "flat" structures emerged.

Such hub structures can be found in two forms: "random" or "rolling." In the first case, arrival and departure times of all flights at a given hub are determined by operational requirements or opportunities. From the connectivity point of view, the resulting structures appear "random." In the case of rolling, the flat overall appearance is achieved by the uninterrupted sequence of directional banks, which maintains clear-cut bank structures in an overall flat pattern. In this section, we will review the conceptual foundations of such "flat" hub structures.

Two distinct hub architectures lead to flat patterns: rolling hubs and continuous hubs.

In *random hub* structures, the timing and sequence of flight movements, inbound as well as outbound, are determined by operational opportunity or requirement in an attempt to optimize aircraft utilization. To that extent, the term "random" hubbing is misleading. Timing of flights is random from the connectivity point of view, but not from the operational perspective. Randomly hubbed structures principally lack

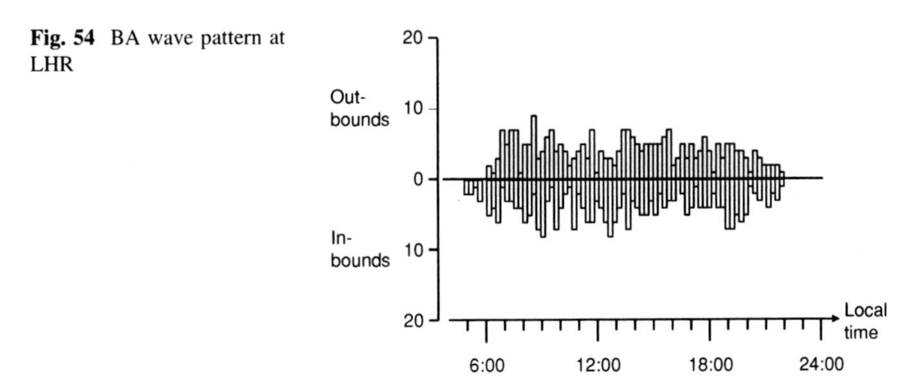

Fig. 54 BA wave pattern at LHR

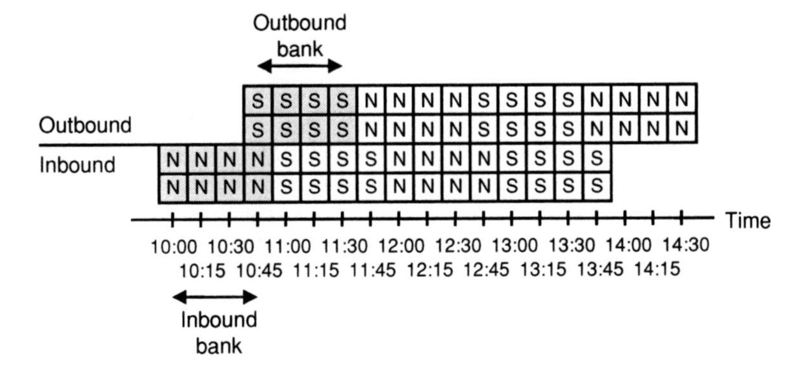

Fig. 55 Banked but nonwaved. Example of a banked but nonwaved hub structure

any temporal structures to support connectivity. As a result, connections happen randomly.

In *rolling hubs*, also referred to as "*continuous hubs*," banks follow each other without any time gap in between. To facilitate high connectivity, rolling hubs maintain clear-cut temporal bank structures but lack temporal "gaps" in between them, so the overall structure appears flat. BA is pursuing the strategy of a focused portfolio of destinations, each served with high frequency and fairly large aircraft. By utilizing the LHR hub, BA can build upon the most powerful catchment area in Europe, but is constrained by a two-runway system. Since connectivity-driven hub structures are not practical for BA at LHR, BA has never tried to implement a waved system at this hub. Figure 54 shows BA's flat hub structure at LHR.

4.1.7 Rolling Hubs: Combining Connectivity and Flat Structures

The conventional pause of activities in between waves is not a conceptual requirement. Distinct banks can follow each other seamlessly, creating a truly flat pattern of flight movements.

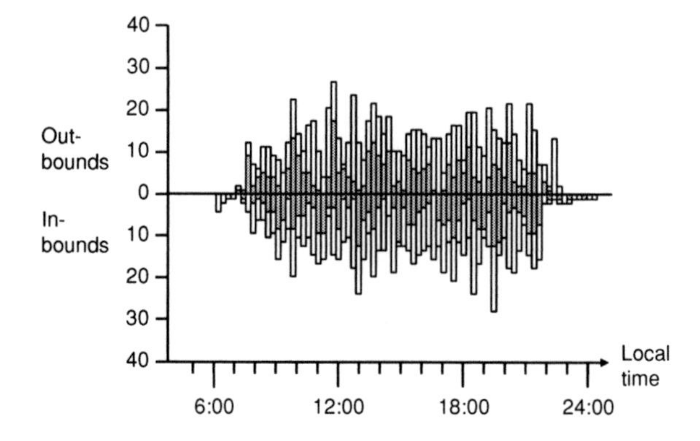

Fig. 56 Rolling hub of AA at DFW 2005. This hub structure is no longer waved and is not yet random. The flights shown in *dark gray* depart to or arrive from easterly directions and are banked as in waved systems. The overall structure, however, is fairly flat

Figure 55 shows a schematic example of a rolling system. Banks of clear North or South directionality follow each other in an alternating sequence and without any temporal gap in between. South outbounds are staggered (with a minor delay to accommodate MCT) atop northern inbound flights. This allows North inbound passengers to quickly connect to South outbound flights. Figure 56 shows the rolling hub structure of AA at its DFW hub in 2005.

In a radical shift away from its traditional wave structures at ORD and DFW, which "has been a hallmark of the deregulated era" (Flint 2002), AA introduced flat schedule structures at both its prime hubs, DFW and ORD. The reasoning was twofold:

- With the advent of Internet-based booking portals that sort by price as the default setting rather than by elapsed time, passenger preferences shifted from preference to short elapsed times to good bargains. With emphasis on elapsed travel time and bank structures being trimmed to the shortest possible connecting times and in spiked structures these structures no longer appeared as the (only) right answer to market expectations.
- Waved structures turned out to be too costly, no longer supporting limited advantages in connectivity.

Due to this paradigm shift for AA's hub at DFW, average connecting times increased by 6 min, reflecting the decline in connectivity. The "average number of connections per arriving flight decreased by 2," while GDS market share data "implied a market share neutral decision," and "on-time performance data implied an improvement in reliability" (Bogusch 2003). Although some commentators misunderstood this departure from waved structures as a departure from hub-and-spoke strategies, the de-peaked structures developed by AA and subsequently DL and others are probably more efficient ways to implement hub-and-spoke, but certainly not an exit from hub-and-spoke.

Zooming into a single North–South bank of the overall structure shown in Fig. 55, we can compare the structural identity of a bank in a rolling bank system with the same structural identity in a waved system.

While rolling hub systems perform well in highly directional hub environments, they are at a disadvantage in hubs with three or more directions. This is due to the decrease of repeat frequency (and the amount of asymmetry in connecting times) of banks with an increasing number of directions. In a two-directional system with inbound and outbound banks taking 45 min each, and adding an MCT of 30 min and a shift of 45 min (see Fig. 55), the average connecting time is 53 min. Each "East-to-West" bank is repeated every 120 min. If a third direction were added, the average connecting time would go up to 82 min, and the repeat time would go up to 180 min. The more directions a given hub is serving, the less adequate a rolling structure becomes, as repeat rates would become too slow.

4.2 Random Hubbing: When Big Beats Complex

By definition, there are no time frames in random structures: connections happen as random incidents between inbound and outbound flights with no apparent structure of arrival or departure times. How many hits are to be expected assuming the random distribution of flights ("if a monkey does the scheduling")? Two types of methodologies prevail to assess the number of expected hits in randomly distributed hub structures:

- Analytical deduction: Based on the original idea proposed by Doganis and Dennis (1989), Dennis (1994) and Dennis (2001), a more advanced algorithm was developed by Jost (2009).
- Experimental analysis by means of Monte Carlo techniques.

4.2.1 Analytical Deduction

For random structures, where inbound flights arrive and outbound flights depart at random times, Doganis and Dennis (1989) have defined the probability to obtain a hit as

$$p = \frac{\text{MaxCT} - \text{MCT}}{T} \tag{18}$$

where T is daily opening hours of the given airport (in minutes).

The actual number of hits to be expected results as:

$$\text{hits} = \text{inb} \times \text{out} \times p \tag{19}$$

Bear in mind the example in Sect. 3.5 in Chap. 3, where the case of 100 inbounds, 100 outbounds, 10 banks, and two directional flows was discussed and led to 500 theoretical hits. Let us again assume the same case, but this time we randomly distribute all flights over a time period from 06:00 to 22:00, equal to 16 h or 960 min opening time. What we get is a "randomly hubbed" pattern. How will connectivity behave in such a randomly hubbed as compared to a waved or banked structure? The bank overlap factor (BOF, see Sect. 3.2.2 in Chap. 3) for randomly hubbed structures is equivalent to formula (18) (Doganis 1989). In our case, MaxCT − MCT may be assumed to result in 75 min as a reasonable figure, and the opening time T to result in 960 min (see above). Hence, (MaxCT − MCT)/T = 0.078125.

Further assuming a strict bi-directional hub structure (making the directional factor 1/2), for the randomly banked structure in our example, we obtain $100 \times 100 \times 0.078125 \times 1/2 \cong 391$ hits:

$$\text{hits} = \text{inb} \times \text{out} \times p \times r \tag{20}$$

When comparing this figure with the equivalent for the waved pattern (500 hits), the latent connectivity disadvantage of random structures becomes apparent.

While this algorithm might work for some standard major hubs in Europe or the United States, it reaches its limit for hubs with significant night-flight activities: The definition of "T" implies a continuous flow of flight movements within a well-defined time window. This formula, however, cannot be applied to airports with fragmented flight clusters with many pauses, or longer pauses, particularly at airports with 24-h passenger flight operations, such as CX at HKG, EK at DXB, or SQ at SIN.

Jost (2009) offers a more generally applicable algorithm to address at least the opening hour problem:

The 24 h of a day are divided into slots of 15 min each; for each such slot, it is determined whether an airport shows flight movements. For any slot without planned movements, the slot is considered closed.

Next, the probability is calculated for the case that a given inbound flight can build a hit with a given outbound flight, constrained by the fact that flight movements may only take place during open slots. This open/close constraint is applied not only to the hub under inspection, but also to the origin airport of the inbound flight and the destination airport of the outbound flight.

The accumulated probabilities equal the number of expected hits.

4.2.2 Monte Carlo Simulation

Another technique, known as Monte Carlo, is more experimental in nature, and as such is closer to the metaphor of the "scheduling monkey" mentioned in Sect. 4.2. Monte Carlo techniques take a given schedule at a given airport and effectively shuffle the timings of all inbound and outbound timings. After each shuffle loop, the resulting number of hits is determined. Following a sufficiently large number of shuffling loops, the resulting average hit score is taken as an indicator of the number of hits to be expected based on random distribution. The advantage of Monte Carlo techniques in this context is that only these techniques permit the far-reaching application of constraints (like auto-adaptive hit windows, see Sect. 2.3.1 in Chap. 2), while the analytical approaches only permit a limited, if not rudimentary, set of constraints.

By relating the observed number of hits to the number of expected hits, one gets a normalized indicator of connectivity sophistication that is not biased by the number of underlying flight movements. It considers far-reaching constraints, and is not receptive to patterns of interrupted periods of activity.

4.3 Who Should, and Should Not, Randomly Hub?

For large hubs, the apparent disadvantage of random hub structures may turn out to be less relevant. Assuming the number of inbound and outbound flights is equal and considering the number of banks as an independent variable, with all else being equal, the maximum number of hits becomes:

$$\text{hits} = \frac{\text{inb}^2}{b} \tag{21}$$

where inb = number of inbound flights (equal to number of outbound flights), b = number of banks.

At large hubs, the squared function in the nominator and the linear proportion in the denominator of formula (21) can over-proportionately compensate for the detrimental impact of an increase in banks (mere linear effect). Not surprisingly, the largest hub worldwide, DL in ATL, was the front-runner in random hubbing.

Let us do some math (with some simplifications applying) on what happened in DL:

In summer 2004, DL operated 11 banks in ATL with 979 inbound and the same number of outbound movements per day, starting from 06:00 to 24:00. DL serves three major directions in ATL, with each direction being able to connect with two others. The duration of each inbound wave is roughly 75 min. For this example, we can safely ignore that in ATL the outbound waves immediately follow the inbound waves or overlap, without any MCT-driven pause in between.

Applying these "waved" parameters to formula (11) results in:

$$\left(979^2 \times \frac{1}{1} \times \frac{1}{11} \times \frac{2}{3}\right) \cong 58,087 \text{ hits} \tag{22}$$

where 979 is the number of inbound movements, assumed to be equal to the number of outbound movements, 1/1 representing the BOF, 1/11 representing the term 1/*banks*, and 2/3 representing the directional reduction factor *r*.

Assuming a flat structure at ATL, MaxCT – MCT to be 75 min, and opening hours to equal 1,080 min, applying formula (20) yields:

$$\left(979^2 \times \frac{75}{1{,}080} \times \frac{2}{3}\right) \cong 44{,}372 \text{ hits} \tag{23}$$

where 979 is the number of flight movements (see above), 75/1,080 is the equivalent of p, and 2/3 represents the directional reduction factor r.

Consequently, we must expect a reduction in the number of hits by about 24% (going from 58,087 down to 44,372 hits). Could increasing the number of flight movements compensate for this detriment? How many more inbound flight movements (assuming number of inbounds = number of outbounds) would be needed to maintain the original 58,087 number of hits? Let us rewrite formula (20) such that we isolate the number of inbound flight movements on one side of the equation:

$$\text{inb} = \sqrt{\frac{\text{hits}}{p \times r}} \tag{24}$$

Applying this formula results in:

$$\sqrt{\frac{58{,}087}{\frac{75}{1{,}080} \times \frac{2}{3}}} = 1{,}120 \text{ inb} \tag{25}$$

To maintain the same level of connectivity, we would need 1,120 inbound movements instead of the original 979. This necessary capacity expansion would be equivalent to 14.4%.

What would happen if we tried to randomly hub a small hub like OS in VIE? In summer 2008, OS operated 6 banks in VIE with a total of about 204 flights inbound and outbound each, starting at 06:00 through 23:00, or a total of 1,020 min opening time. OS mainly serves East–West flows. The duration of each inbound bank is roughly 60 min. Again, we ignore that in VIE the outbound wave immediately follows the inbound wave, without any MCT-driven pause in between.

The theoretical number of hits, according to formula (24), is calculated as:

$$204^2 \times \frac{1}{1} \times \frac{1}{6} \times \frac{1}{2} = 3{,}468 \text{ hits} \tag{26}$$

If we convert the six-bank system into a random pattern with all flights being equally distributed over the same operating hours, assuming MaxCT − MCT to be 75 min, then we get:

$$204^2 \times \frac{75}{1{,}020} \times \frac{1}{2} = 1{,}530 \text{ hits} \tag{27}$$

Consequently, we have to expect a reduction in the number of hits by roughly 56% (going from 3,468 hits down to 1,530). In the case of OS at VIE, how many more inbound flight movements would be needed to maintain the original 3,468 number of hits? By applying formula (24), we can assess the number of required movements to maintain the original 3,468 hits:

$$\sqrt{\frac{3{,}468}{\frac{75}{1{,}020} \times \frac{1}{2}}} \cong 307 \text{ hits} \tag{28}$$

As a result, we would need a total of 307 movements in a random scenario for OS at VIE, as compared to the original 204 movements in a waved structure. This means that to maintain the connectivity offered in a waved system at VIE, OS would have to expand its flight movement capacity by 50%. In essence, neither a connectivity drop of 56% nor a required capacity expansion of 50% appears feasible or reasonable.

Alternatively, OS could opt to accept a much wider hit window, resulting in many more, fairly slow connections. This would mean that 204 flights equally distributed over six banks and 1,020 min of operating hours (as in the waved system) would result in 1,020/6 = 170 min duration per inbound bank. If we combine an inbound bank of 170 min with a corresponding outbound bank of 170 min, the slowest yet permissible connection would connect the first inbound flight with the last outbound flight, which departs 2 × 170 = 340 min later. For comparison, the slowest permissible connection in the current waved system of 60 min bank duration is 2 × 60 = 120 min. All three issues would be prohibitive: (1) the slowdown to 340 min, (2) the otherwise necessary capacity expansion, or (3) the drop in connectivity. The comparison of mega-hub ATL with hublet VIE shows the overwhelming importance of scale on the applicability of random hubbing.

In essence, random hubbing is a nonoption for small or medium-sized hubs, which depend on competitive scores of connectivity and swift connections. Random hubbing is an option only for very large hubs. Due to the likely need to compensate part of the hit loss when introducing random hubbing by capacity expansions, random hubbing must assume a sustainable growth perspective.

4.4 Is Network Strategy an Experimental Discipline? On Hub Structure Evolution

With DL's hub at ATL, one of the most prominent flat hub structures recently reverted back to a profoundly waved pattern. Figure 57 shows the evolution of DL's flat hub structure at ATL in 2005, succeeded by a spiked pattern. While the reasoning behind this is speculative, the introduction of efficiency-driven flat structures was decided upon during an industrial downturn following 9/11, and DL's return to waves was decided upon at a time of strong growth.

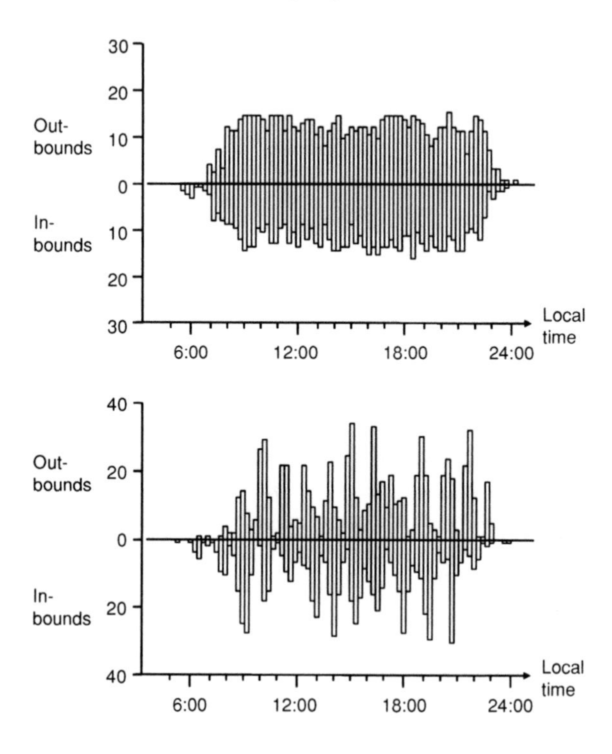

Fig. 57 The evolution of DL's hub structure at ATL from flat (2005: *top*) to waved (2009: *bottom*)

4.5 Special Topic: Micro Banks

In Sect. 4.1.7, we see that in conventional rolling hubs, due to the extended durations of typical inbound and outbound banks in rolling structures, feasible connecting times and the repeat rates of banks easily become prohibitive, particularly when more than two or three directional flows must be served at a given airport. This disadvantage could be overcome by leveraging short bank durations rather than lengthy ones. In Fig. 55, the duration of banks was assumed to total 45 min, reflecting industry standards for rolling hub systems at the time they were in vogue. In Fig. 58, we use the same logic but assume each bank to last only 15 min. Let us further assume the same parameters as in the example used for Fig. 55: 16 inbounds from the North and South each, mirrored numbers on the outbound side, and an MCT of 30 min. The following connection and cycle repeat times would result:

NS flows take 30 min to connect, as do SN flows. If a passenger misses a connection, the respective directional flow is repeated as soon as possible after 30 min. Hence, delayed passengers could be accommodated with a fast recapture, if necessary, at least for destinations served by each bank (or "pillar flights," see Fig. 60). In the context of rolling hub structures, we refer to such short banks as micro banks, which are similar to rapid banks (see Sect. 3.3 in Chap. 3). Rapid banks refer to nonflat structures, and micro banks refer to flat rolling hubs. The

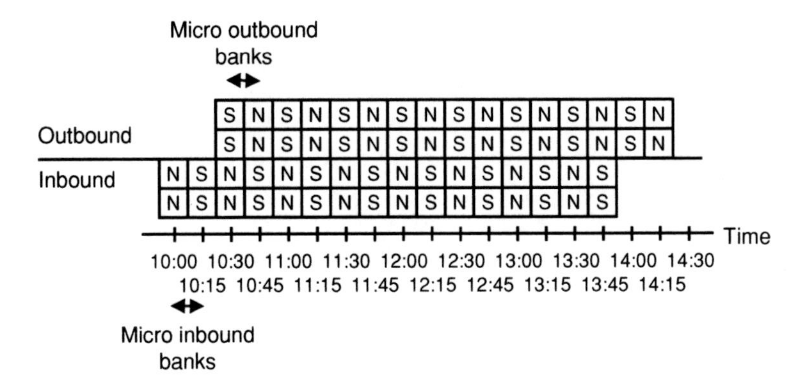

Fig. 58 Micro banks in a two-directional framework

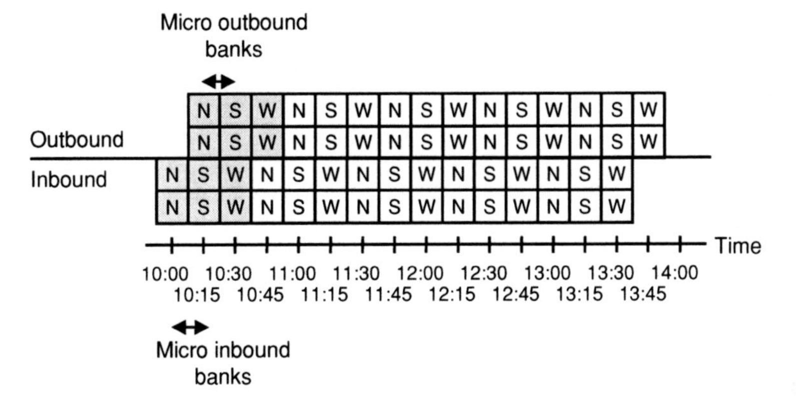

Fig. 59 A three-directional micro bank scenario

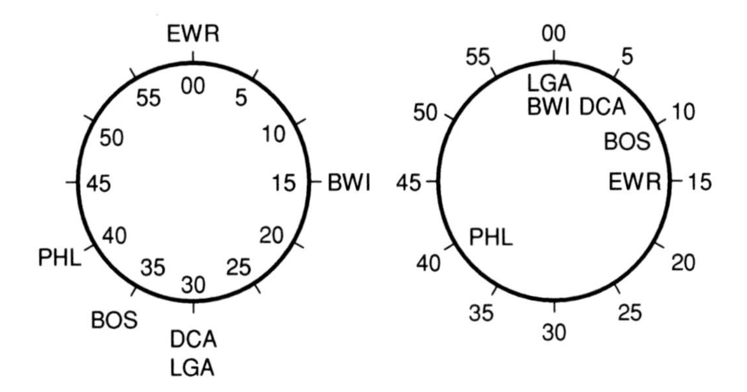

Fig. 60 Departure and arrival times of DL pillar flights (summer 2005). On the *left side*, departure times of flights to ATL are shown; on the *right side*, departure times at ATL are shown

advantage of micro banks over rapid banks is that micro banks provide faster repeat cycles than do rapid banks. The advantage of micro banks over random structures is that after MCT completion, the respective hit window immediately starts with flights headed in a direction that can be connected. In random structures, the probability of running into a flight back to N at 10:30 (after arriving from N at 10:00) would be 50%. In micro banks, this probability is lowered to 0%.

Of course, the duration of micro banks can be further reduced to slices of 5 min per micro bank. It is important to understand that the advantages of micro banks assume that all directions are served by the same fleet. If directions are highly fleet specific, the advantages of micro banks are diluted.

The optimum duration of a micro inbound or outbound bank is given by:

$$\mathrm{BDI} = \mathrm{BDO} = \frac{\mathrm{MCT}}{\mathrm{DR}} \tag{29}$$

Hence, the more directions (DR) are to be accommodated, the thinner the micro bank slices have to be cut, or the less adaptive micro banks are as a structure to accommodate the needs of an omnidirectional hub. In the United States, most hubs are either two-directional (DFW, ORD) or three-directional (ATL). Figure 59 gives a schematic example of a three-directional micro bank structure with an MCT of 30 min.

In this example, NS flows connect after 30 min, and SN flows after 45 min; SW flows connect after 30 min, and WS flows after 45 min; WN takes 30 min, and NW takes 45 min. The repeat time is 45 min. Note that the connecting times for direction and counter-direction are asymmetric; the same holds true for rapid banks. The more directions that are to be served in micro or rapid bank frameworks, the more asymmetric the resulting connecting times become.

Micro banks can easily become very intricate or prohibitively complex when:

More than three directions must be served.
Directional segments carry very different volumes of traffic.
A significant number of long-haul flights must be accommodated, both inbound and outbound.
Few destinations can be served with high frequency.

4.6 Not for the Fainthearted: Low-Cost Long Haul

Various experiments have been conducted to implement the low-cost concept to long-haul markets, but most have failed. The reasons are:

Few cost advantages of short/medium-haul LCCs can be applied to long haul. One obvious cost advantage of the LCC business model, however, can immediately be applied: tight seating. This results in many more seats per aircraft, or a better exploitation of "earning capacity." On long haul, most flight variable costs are either regulated or monopolized, or both, leaving little room for cost advantages. However, flight variable costs mark a much higher share of overall costs for long-haul operations than for short/medium haul. As a result, the cost advantage of short/medium-haul diminishes with longer hauls.

Aside from tight seating, the only effective driver of cost advantages for long-haul low-cost operations are passenger variable costs. Hence, "de-frilling" service concepts is the only cost advantage not available to full-service carriers on long haul. On long haul, however, service becomes a significantly more important issue than on shorter trips. Aside from the structural cost advantages of the "de-frilled" business model, long-haul LCCs may benefit from the blank-sheet advantage of a startup to negotiate labor and airport service contracts, giving them an additional cost advantage over the incumbents.

Full-service carriers have also learned to strike back: The seats in the back end of a wide-body long-haul aircraft can be marketed at marginal costs—at the same level of fares that the long-haul LCCs market as average fares. In this case, the better service concept of full-service carriers makes the difference, to the disadvantage of long-haul LCCs. Some recent LCC long-haul new entrants strongly emphasize the advantages of the blank-sheet approach combined with dense seating. Initial indicators of success appear promising, but more time is needed to assess the sustainability and general applicability of this model.

4.7 Breaking the Rules of Network Structure? On Pillar Flights

Within the family of random hub structures, pillared structures differ from a fully continuous, commercially random flow of flight movements. Pillared systems consist of the flights of highest commercial importance (hence "pillars") that are timed to depart or arrive at (for instance) each full hour and half-hour. All other flight movements are then randomly placed around these pillar flights. Such pillar

Table 1 Structural characteristics of various hub architectures

	Random	Banked
Flat	Flights are evenly distributed ("de-peaked") over the course of the day without any apparent temporal structure Example: DL at ATL, 2005	Flights are evenly distributed ("de-peaked") over the course of the day, building upon seamless sequences of temporal clusters of flights of a particular nature, such as direction or value Example: AA at DFW, 2005
Waved	–	Sequences of temporal clusters of inbound and outbound banks of flights, separated from each other by periods of reduced activity Example: AF at CDG, 2009

flights were prominent in the first release of DL's revised hub system at ATL in 2005 (see Fig. 60).

Table 1 summarizes the structural characteristics of various hub architectures.

Chapter 5
Case Studies

Abstract In this chapter, we will discuss two case studies: one generic bottom-up case provides an objective, adds some constraints, and then starts with a blank sheet. The other case examines the top-down redesign of KL's hub for the winter schedule 2009–2010.

5.1 Case #1

We will develop this case step by step. Let us assume a new bank needs to be built for a small hub with eight inbound and eight outbound flights, all flights are of comparable short/medium-haul length, all connections are of similar value, all flights connect with an MCT of 30 min, and all aircraft can turn around with a TAT of 30 min. Furthermore, all flights can be efficiently rotated at the respective outstations. In the simplest case, the theoretical maximum of possible connections would be $8 \times 8 = 64$ hits. This would require all eight inbounds to arrive at the same time, and all eight outbounds to depart at the same time. Let us further assume a capacity-constrained airport with a maximum of four inbound and four outbound movements per 15-min slot. The scheduler faces a dilemma: whether to optimize connectivity or aircraft utilization. Figure 61 shows the connectivity-driven option, while Fig. 62 illustrates the corresponding utilization-driven option for structuring the bank.

The resulting figures for the number of hits, average connecting time, and average idle time of aircraft are shown in Table 2.

Table 3 shows the performance indicators for the utilization-driven option.

Comparing the performance figures of both options, the utilization-driven bank structure clearly provides 25 percent fewer hits (48 instead of 64); however, these hits are almost 25 percent faster on average (35 min instead of 45 min). Most importantly, the utilization-driven structure can be produced with zero idle time: all aircraft turn around at a minimum TAT of 30 min. In Sect. 3.2.2 (Chap. 3),

P. Goedeking, *Networks in Aviation*, DOI: 10.1007/978-3-642-13764-8_5,

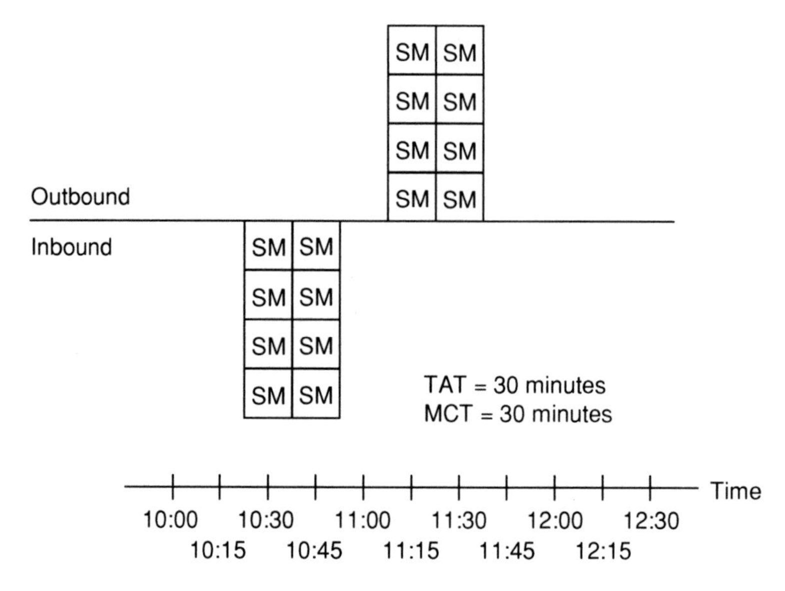

Fig. 61 A connectivity-driven solution to the short/medium-haul flights in case #1

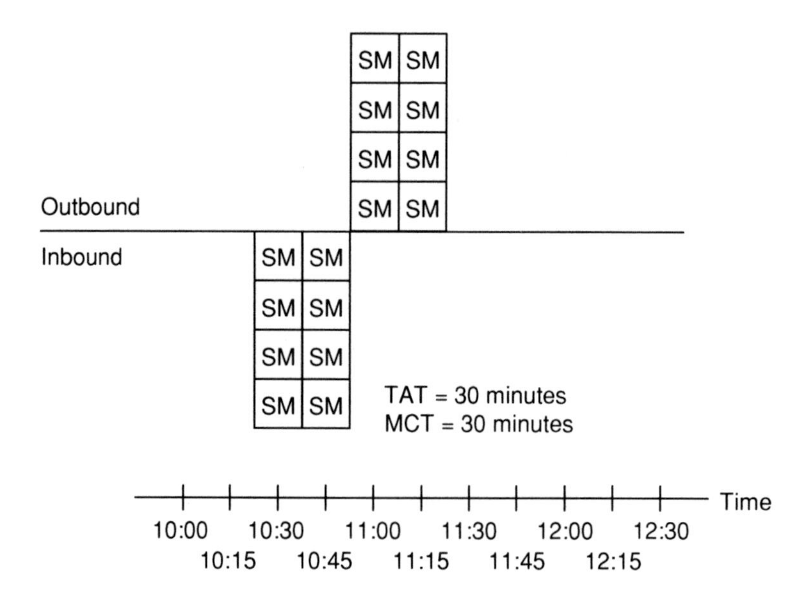

Fig. 62 A utilization-driven solution to the short/medium-haul flights in case #1

we discussed the impact of overlapping inbound and outbound banks. This simple case makes it even more evident. Are the advantages of aircraft utilization and speedy hits worth the compromise in the absolute number of hits? If the respective competitive environment requires fast connections as a precondition to attract

Table 2 Performance of the connectivity-driven solution shown in Fig. 61

Number of hits	64
Average connecting time	45 min
Average idle time	15 min

Table 3 Performance of the utilization-driven solution shown in Fig. 62

Number of hits	48
Average connecting time	35 min
Average idle time	0 min

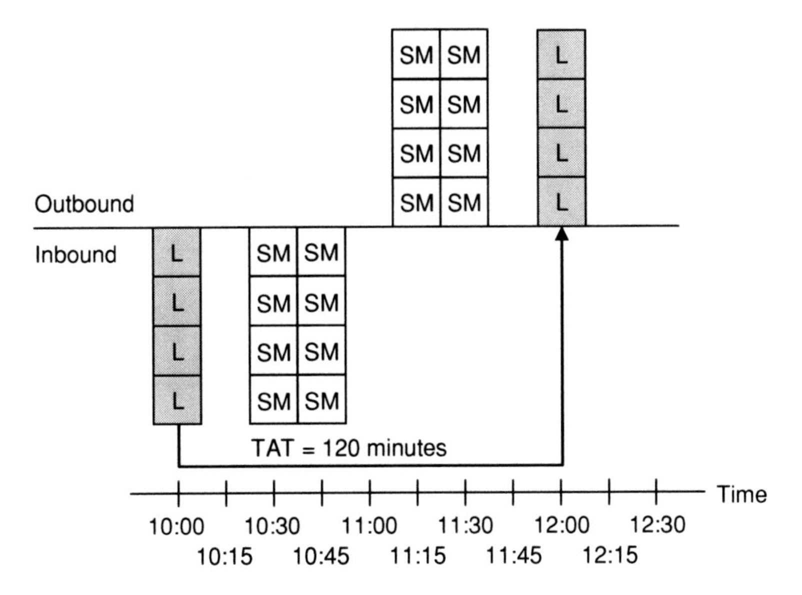

Fig. 63 The short/medium-haul pattern in Fig. 61 enriched with long-haul flights

sufficient traffic, banks should not overlap, and compromises in regard to aircraft or asset utilization must be accepted. If competitive pressure affects cost more than connectivity, banks are more likely to overlap. LH's hub in FRA (see Fig. 31) illustrates the latter, while NW's hub in DTW (see Fig. 26) demonstrates the former.

Now let us assume that we add four inbound and four outbound long-haul flights, with a TAT of 120 min and an MCT for long haul to short/medium haul (or vice versa) of 60 min. We examine three structural configurations to bring together the short/medium-haul flights with the long-haul flights. The first option (see Fig. 63) positions the long-haul flights around the structural pattern developed in Fig. 61.

The corresponding performance indicators are as follows (Table 4).

Another alternative would be to build the long-haul flights around the productivity-driven pattern (see Fig. 64) developed in Fig. 62.

Table 4 Performance indicators of the structure shown in Fig. 63

Short/medium haul to short/medium haul or vice versa	Number of hits	64
	Average connecting time	45 min
	Average idle time	15 min
Long haul to short/medium haul or vice versa	Number of hits	64
	Average connecting time	83 min
	Average idle time	0 min

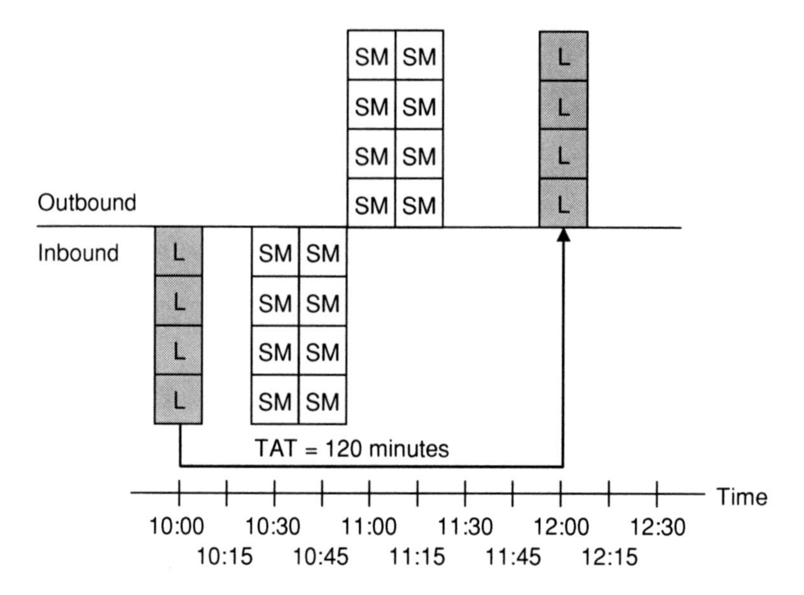

Fig. 64 The productivity-driven case in Fig. 62 enriched with long-haul flights

Table 5 Performance indicators for the productivity-driven case shown in Fig. 64, enriched with long-haul flights

Short/medium haul to short/medium haul or vice versa	Number of hits	48
	Average connecting time	35 min
	Average idle time	0 min
Long haul to short/medium haul or vice versa	Number of hits	64
	Average connecting time	75 min
	Average idle time	0 min

The following performance indicators would result (Table 5).

Another approach to the same problem would be the following: instead of using short/medium-haul flights as a starting point, this approach uses the long-haul flights. Let us assume a group of four inbound long-haul flights, all arriving at 10:00 local time (see Fig. 65). Given the long-haul MCT of 60 min, the outbound bank that de-feeds this long-haul inbound bank commences at 11:00.

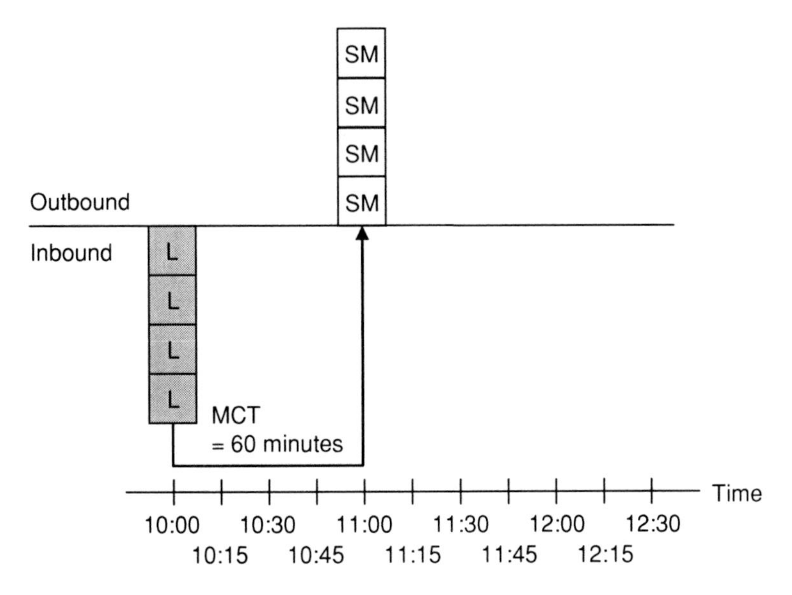

Fig. 65 Long haul and MCT. The long-haul inbound and related MCT dictate the timing of the corresponding outbound de-feeder bank

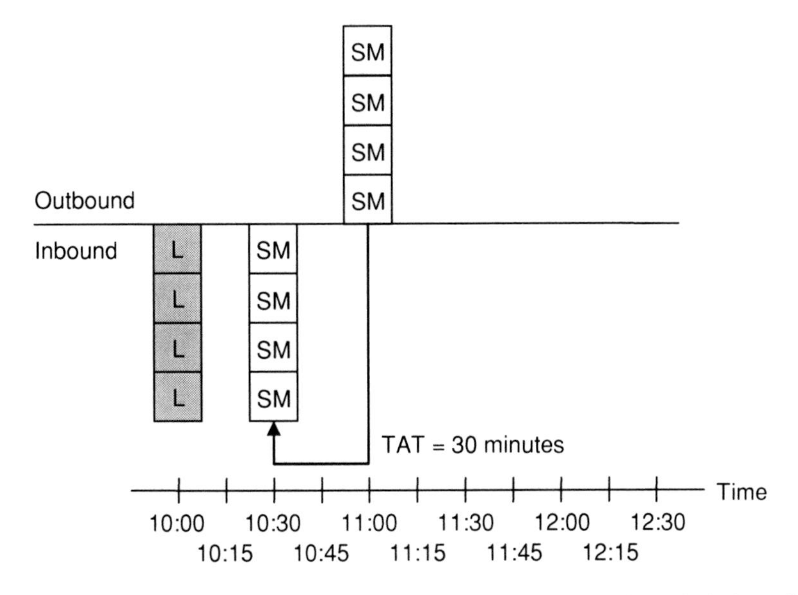

Fig. 66 Short/medium haul and MCT. The 30-min TAT determines the arrival time of the aircraft forming the de-feeder bank (at 11:00) of the long-haul inbound bank (at 10:00)

Given the objective of optimum aircraft utilization, when do the 11:00 de-feeder bank aircraft need to arrive? With a 30-min TAT for the feeder/de-feeder aircraft, the aircraft must arrive at exactly 10:30 (see Fig. 66).

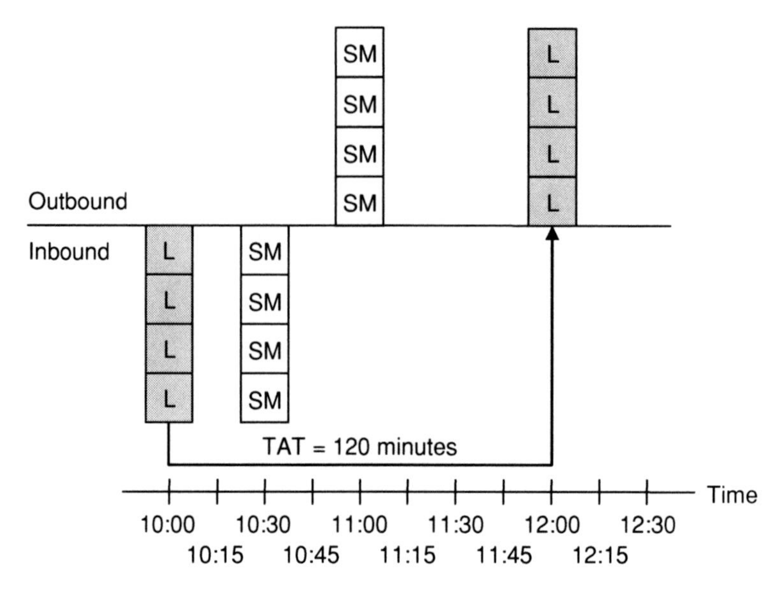

Fig. 67 Long-haul to long-haul connections. The outbound long-haul bank follows the inbound long-haul bank with a delay of 120 min (long-haul TAT)

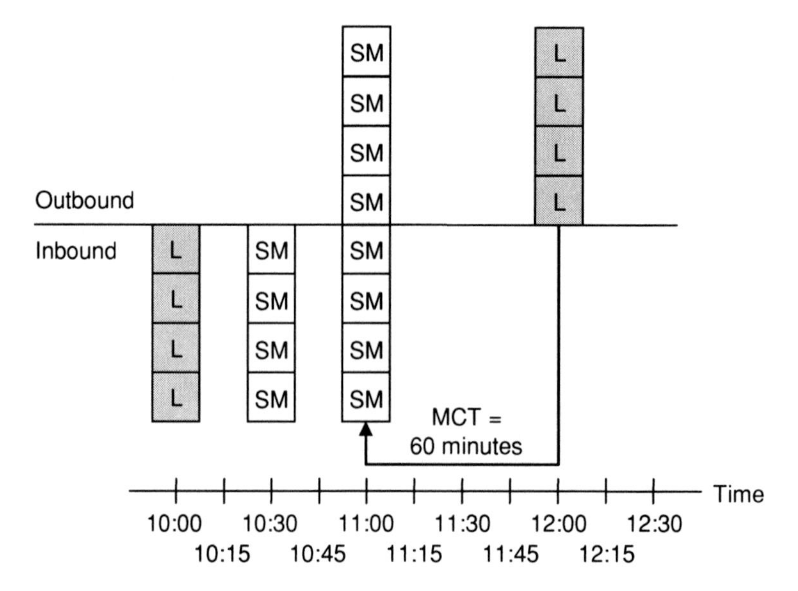

Fig. 68 Short/medium haul to long haul. The bank feeding long-haul outbound flights starts 60 min before the long-haul outbound bank, or 11:00

As a next step, we need to assess the right departing time for the long-haul flights that arrive at 10:00. Assuming a 120-min TAT for long-haul aircraft at their base and any applicable MCT to outbound long-haul aircraft being shorter, the

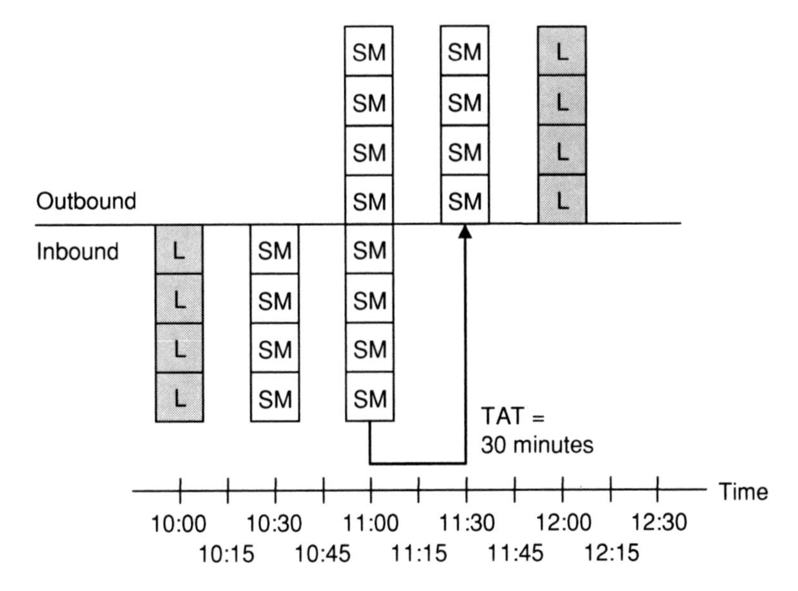

Fig. 69 Short/medium haul to short/medium haul. This schedule scenario adds the departure bank of the long-haul feeder bank by applying a 30-min TAT

outbound long-haul bank must follow the long-haul inbound bank by 120 min; this results in an absolute time of 12:00 (see Fig. 67).

The feeder flights for the outbound long-haul bank, commencing at 12:00, must start with a lead time given by the MCT for short/medium-haul to long-haul flights, in this case assumed to be 60 min. Hence, the bank feeding the long-haul outbound bank starts at 12:00 – 60 min = 11:00 (see Fig. 68).

Finally, we must determine the departure time for the short/medium-haul feeder flights added in Fig. 68. As in Fig. 66, adding a 30-min TAT will result in the productivity-optimized departure time of the corresponding outbound bank at 11:30 (see Fig. 69).

The resulting performance scores are summarized as follows (Table 6).

By comparing three ways to combine long haul and short haul, we can clearly see that there is no "best" solution. Each approach has its specific strengths and weaknesses: the option built around the connectivity-driven short/medium-haul bank offers the most, but relatively slowest, hits. The other two options—one

Table 6 Performance figures for the schedule scenario shown in Fig. 69

Short/medium haul to short/medium haul or vice versa	Number of hits	48
	Average connecting time	40 min
	Average idle time	0 min
Long haul to short/medium haul or vice versa	Number of hits	64
	Average connecting time	75 min
	Average idle time	0 min

based on a productivity-dominated bank, and the other developed around the needs of long-haul flights—are similar in overall performance, but differ in structure. The option based on a productivity-dominated bank design has a slight advantage in terms of connecting times.

For all three options, we can apply two generic rules:

- Structure follows strategy. What kind of performance is desired? What kind of structure best helps to achieve such performance?
- To some extent, schedule structure compromises asset productivity, the number of hits, and performance time.

5.2 Case #2

Beginning with the winter 2009–2010 season, KL introduced a new hub structure for its AMS hub. This change in hub structure provides a good example for studying a real-world case top-down. Figure 70 compares the "old" structure, built upon six waves, with the "new" structure, reduced to five waves.

In Fig. 70, we can observe the:

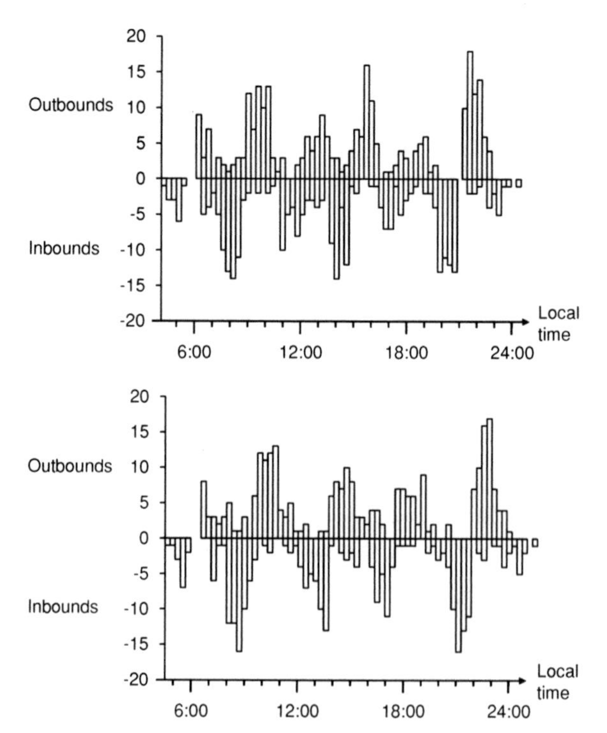

Fig. 70 KL at AMS. *Top*: the structure before the 2009–2010 winter schedule. *Bottom*: the structure from the 2009–2010 winter schedule going forward

Table 7 Performance comparison of the "old" versus "new" hub structures of KL at AMS

	Indicator	Old	New	Percent change (%)
Capacity	Number of destinations	245	221	−10
	Number of movements	7,366	6,543	−11
	Seat capacity	1,056	953	−10
	Available seat kilometers (ASK)	3,119	2,909	−7
	Average frequency/route	16.3	16.2	−1
Connectivity	Connections	55,646	53,296	−4
	Connections per inbound	15.1	16.3	8
	Hit feed per long-haul outbound	21.8	20.0	−8

- *Reduced number of waves.* The three waves between 11:00 and 19:00 (local time) have been condensed to two waves. As shown in Sect. 3.4 (Chap. 3), a reduction of banks should lead to a significant increase in the number of hits.
- *Reduced overlap.* While the three former waves overlapped significantly (inbound-outbound overlap), the two successor waves do not. Reduced inbound-outbound overlap could lead to more, but slower, hits (see Sect. 3.2.2 in Chap. 3).
- *Tighter banks.* The durations of the inbound and outbound banks (BDI, BDO) in the new waves are reduced as compared to their predecessors. Such a measure is expected to reduce the connecting time of hits (see Sect. 3.2 in Chap. 3).

A quantitative analysis of this new bank design reveals that KL has reduced capacity by around 10 percent with its new schedule (Table 7). A reduced capacity should over-proportionally reduce the number of hits (see Sect. 3.1 in Chap. 3). However, the number of hits in the new system went down by only 4 percent, or under-proportionally. When adjusted to the number of underlying inbound flight movements (the passenger perspective), the average number of hits per inbound went up by 8 percent. KL has countered the disadvantage of lowered capacity by reducing the number of waves, eliminating bank overlap, and tightening the banks.

Chapter 6
Planning and Controlling Networks: Networking Higher Math, Intuition, and Power Play

Abstract In this chapter, we review the key organizational, structural, and procedural options for network management, the basic fundamentals of network control, and the most important IT tools for supporting such processes.

Despite an endless number of variations, the network planning process can be divided into three major phases (see Fig. 71). The first phase takes the longer-term or strategic perspective, while the second phase focuses on the medium-term or next-season schedule. The third phase addresses the short-term or ongoing schedule season.

In network planning, a schedule structure evolves incrementally as the scheduler adds frequencies or destinations, aligns fleet assignments, improves code shares, and optimizes time-of-day or day-of-week patterns. Zero-based network or hub redesign, in contrast, is rarely used. However, longer-term or strategic network planning incorporates many facets of fundamental network and hub redesign. This type of planning should be routinely checked against the zero-based redesign in order to confirm the sustainability of the current network system.

6.1 The Longer-Term Perspective: Strategic Planning

In the past, many airlines considered that longer-term or strategic planning entailed the same level of detail as scheduling for the respective next season. The required fleet was determined by a fully constrained rotational evaluation of a long-term plan that reflected market opportunities. Today, the process is reversed: The planned evolution of assets (aircraft and infrastructure capacity) is provided as strategic input to network planning, which in turn plans the product and production accordingly. Apart from financial planning requirements, the key reason for this change was that longer-term planning barely resembled what was finally implemented. Therefore, there was no value in adding extensive details to long-term

P. Goedeking, *Networks in Aviation*, DOI: 10.1007/978-3-642-13764-8_6,

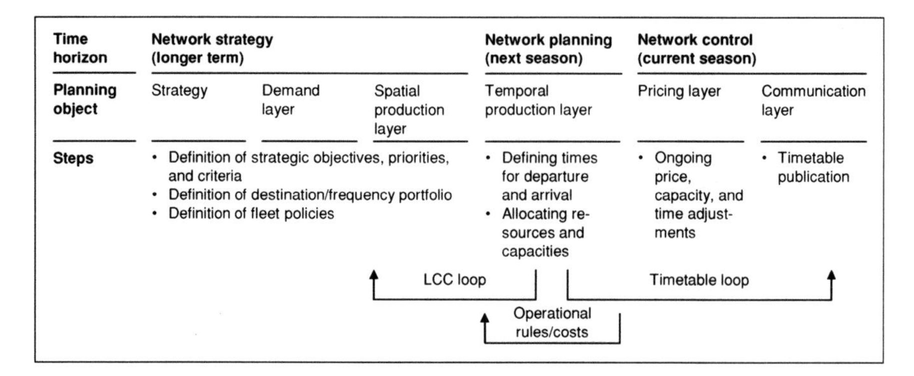

Fig. 71 The phases of zero-based hub redesign. The comments under "Planning objective" refer to the discussions in Chap. 10

plans. Additionally, the prevalence of operational detail in long-term planning often drew attention away from the truly important aspects of competitive benchmarking: the need to constantly challenge basic strategic beliefs and maintain market adaptability.

When planning network strategy, it is good practice to view this phase as a zero-based planning process. That way, external market forces and internal capabilities can be constantly weighed against established network structures. Accordingly, we will describe this process as "zero based."

The typical airline portfolio of destinations is rarely the immediate result of thorough market research. Other factors, such as the history of its evolution, its network, and its production requirements, all shape the eventual portfolio of destinations and related frequencies. A good example is offered by pre-liberalization airlines in Europe, where each nation operated its government-owned "flag carrier." The political considerations of offering flight services to specific areas of the respective nations resulted in different network structures, as well as different operational and financial performance figures.

Even in today's liberalized European market, many remote regions cannot operate without subsidized air services. In Europe, the European Union (EU) Commission permits state aid to serve certain destinations. This principle is called "Public Service Obligation (PSO)," and its applicability is highly constrained and regulated. Without such PSOs, many islands in Greece and Portugal would be disconnected from the world, as would remote regions in Ireland and Finland. PSOs may significantly determine the overall scope of an airline's network.

6.2 The Medium-Term Perspective: Tactical Planning

When planning medium term [similar to planning the next season, or "tactical planning" (Klein and Steinhardt 2008)], either the current schedule or an approved strategic scenario is taken as baseline. This baseline is then incrementally aligned to meet the product and production requirements of next season's plan.

Invariably, this alignment is a process of frequent iterations, particularly between network planning and operational departments (crew, fleet, maintenance). However, these iterations often turn into organizational conflicts. While commercial objectives naturally differ from operational aims, and some dispute is natural and desirable, the cause for escalating conflicts at these organizational interfaces generally is due to deficient process organization rather than to inappropriate individual behavior.

Often, the underlying workflows foresee "approval" feedback loops from operational to commercial departments, but lack structured and well-defined feed-forward loops of the same direction. Network planning needs to know the rules and criteria applied by operations to evaluate schedule scenarios: What makes a schedule operationally expensive or efficient, vulnerable or robust? Schedules forwarded to operations for evaluation must transparently comply with these rules and criteria. Network planning is the interface between organizational responsibility and accountability for balancing planned revenues and (direct) operational costs (see Sect. 6.6).

In particular, the complexity of network planning has resulted in the development and deployment of complex IT tools that support these processes, as well as the capacity to control planning complexity at ever-higher levels. Airline profitability models (APMs) are one prominent example. Using a schedule scenario and the schedules of competitors as input, these tools generate estimates of likely market shares per O&D or route. These functionalities are variations of the market share models discussed in Sect. 2.3.4 in Chap. 2. As a next step, these tools multiply the expected market shares per O&D by the estimates of total passenger demand as per O&D (see Sect. 1.8 in Chap. 1) and by relevant unit production costs. A forecast of the overall profitability of the underlying schedule scenario is then provided.

APM tools first emerged in the late 1980s and early 1990s and quickly became the standard central network planning device for all airlines. The assessment of likely market shares soon suffered from the inability of earlier versions to incorporate fare differences. On markets with heavy competition between network carriers and LCCs, APMs tended to fail—and still do—despite advanced APM versions that incorporate pricing parameters. Such tools are now used to support, not substitute, professional judgment.

Another disadvantage of APMs is that multiplying market share estimates with passenger demand estimates to yield likely revenues often results in wide margins of error. Similarly, applying unit costs to convert network cost drivers into expected network costs runs the risk of extrapolating limited productivity. A more highly efficient network scenario would be hampered by applying unit costs stemming from older, less efficient network versions. As a result, "profitability" projections based on doubtful revenue and cost projections did not contribute to the credibility of APM tools. The weakness of APM tools is rooted in their claim to provide "exact" numbers; in reality, they provide vague estimates that carry a significant margin of error.

However, APM tools are indispensable in comparing schedule scenarios. Here, the error of data or calibration is the same in all scenarios explored, and the resulting ranking of scenarios should be reliable, even if they are not the absolute scores.

The final step in network planning is preparing, conducting, and managing the outcome of IATA slot conferences. Network plans require different or new slots. However, such slots may not be available at highly congested and IATA-coordinated airports, and may require high levels of flexibility for slot-sensitive schedule scenarios. Sophisticated slot management tools help schedulers to quickly adapt schedule scenarios to the availability of particular slots. At some airports, notably LHR, a gray market of slot trading has emerged. As a result, the emergence of markets for slots is enabling the market-driven valuation of slots. So far, slot values are highly volatile. In the recent past, however, some slots achieved prices in the order of double-digit million EUR.[6]

6.3 Short-Term and Ongoing Perspective: Network Control

During the ongoing season, the network schedule is subject to constant changes in timing, equipment, frequency, capacities, and prices. [This phase is also referred to as "operational planning" (Klein and Steinhardt 2008)]. An ongoing season, in particular the few days before the day of operations, is subject to the complex and sophisticated concepts, tools, and procedures of revenue management. Due to the enormous impact on profitability, airlines seek to align every aspect of capacity, timing, and fares to the short-term evolution of O&D specific demand and competitive environment. The need to constantly align has put great pressure on the extensive lead times of planning procedures, such as fleet and crew assignment. Five years ago, respective lead times for major network carriers were in the order of months, but advanced airlines have since reduced this figure to days. It should be noted that the ability to align types of aircraft at short notice comes at the expense of high complexity costs, which in turn are difficult to assess. For that reason, some airlines focus on sophisticated nesting control (see below) rather than sophisticated short-term fleet assignment.

One key element of revenue management is the concept of "nesting" (Figs. 72, 73): A typical network carrier today offers more than a dozen booking classes on each flight. Each booking class varies in regard to certain service and flexibility privileges. For the most expensive booking classes (fare types), few if any constraints apply. For the lower-priced booking classes, passengers must accept certain restrictions such as the famous "Sunday excursion" rule (a complete Sunday must

[6] This raises the question of who owns slots. From a legal point of view, this question is unanswerable. If slot values were the property of airlines (as suggested by the fact of trading slots among airlines), they would have to be activated in airline balance sheets. Even though slot rights are substantial and real, they are not activated today.

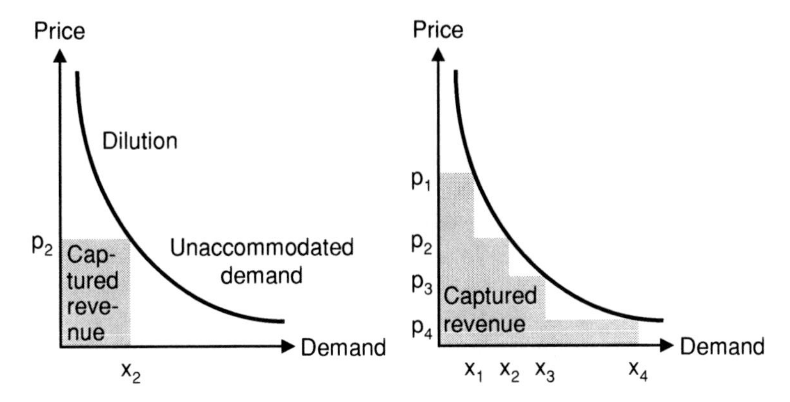

Fig. 72 From a single booking class (*left*) to multiple booking classes (*right*) to optimize captured revenue

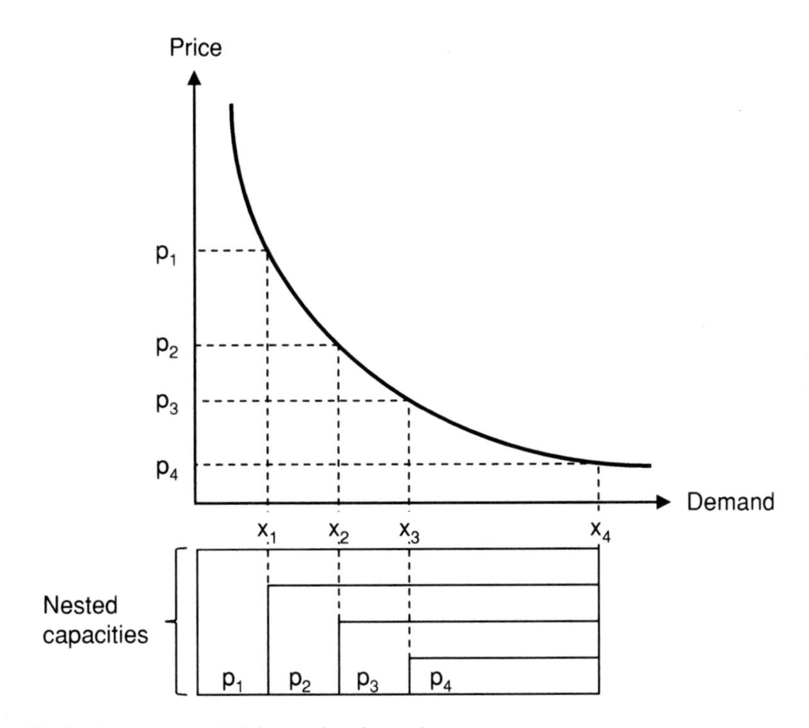

Fig. 73 Nesting to protect highest value demand

be accepted between the outbound and return flight), making these booking classes less attractive for passengers who frequently fly on business trips. In principle, nesting means that 100 percent of the seat capacity of a given flight is available for booking at highest fares. *Of that capacity*, a few seats *also* are available for the low-priced fares. If the entire capacity can be sold early onto high-fare passengers, the high-value bookings automatically overwrite the capacity "nested" for lower-value passengers. If all capacity could be sold early on at low fares, access to capacity beyond the nested capacity (just a few seats) would be blocked. If seat capacity were fixed rather than nested, demand for high-value booking classes would have to be turned away to accommodate lower-value demand.

Nesting permits the use of full-range price elasticity. A single fare or price point corresponds to just one level of the "willingness to pay" this price and excludes higher levels of price elasticity. Consequently, a multitude of price points for the same product (a flight from A to B) makes the most sense. However, passengers who are willing to pay 500 euros for a ticket, but who could get a seat on the same flight with the same level of service for 100 euros, would naturally avoid the higher fare and take advantage of the better bargain. If the "downtrading" to the 100 euro fare required accepting a Sunday excursion rule and being 60 years of age or older, the typical business traveler would have to refrain from such bargains. Therefore, those "restrictions" applying to bargain fares are called "fences," as they are designed to fence off higher-value demand from access to bargain tariffs (Fig. 74). A science in itself, fencing has triggered surprising creativity among price planners (see Appendix, "Painting and pricing").

The nesting principle carries the risk that passengers may not show up to book their reserved seats. In this case, it would be better to open the capacity for lower-value demand in advance. Managing this risk requires forecasting the day-specific evolution of demand per O&D and booking class.

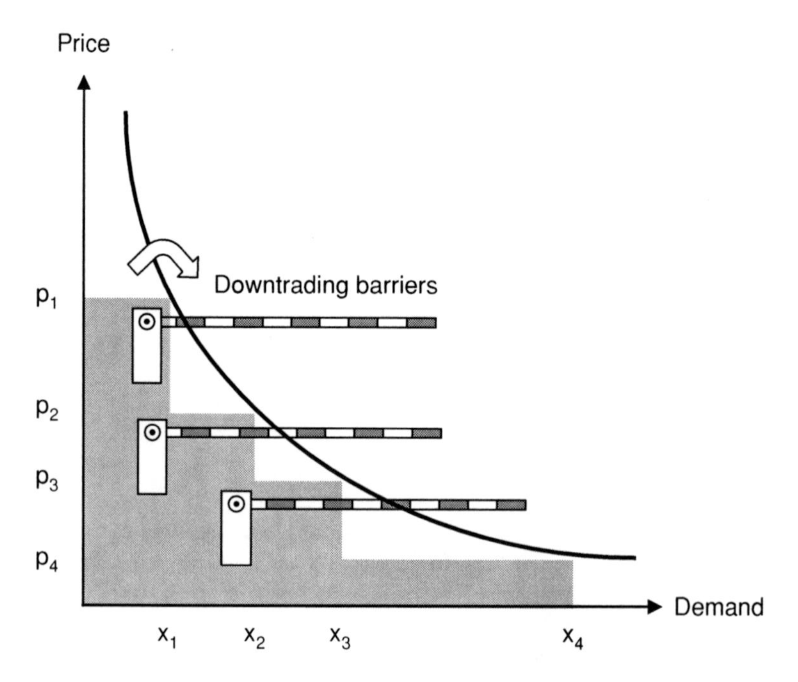

Fig. 74 Fencing to prevent downtrading

6.4 Organizational Structures of Network Management: How to Balance Power

Before deciding upon meaningful organizational structures or processes, the role of network management within an airline must be defined. The key question is: Who is responsible and accountable for revenues and costs? Two prototypical models prevail.

In the first model (see Fig. 75), network responsibility rests within a broader commercial unit, most frequently in parallel to sales, and is designed to act as a revenue center. This commercial unit is a fully accountable revenue center, on par with a likewise accountable operational cost center. In this framework, rigid procedures are needed to ensure the proper balancing of revenues and costs.

In the second model (see Fig. 76), network management is positioned as a profit center in between sales and operations. This way, the responsibility and capacity to optimally balance cost and revenue rest with network management. One key issue to consider: In airline organizational setups of this kind, network management units become powerful.

6.5 Structures Within Network Management Organizations

Most network carriers follow a variation of the following three organizational paradigms.

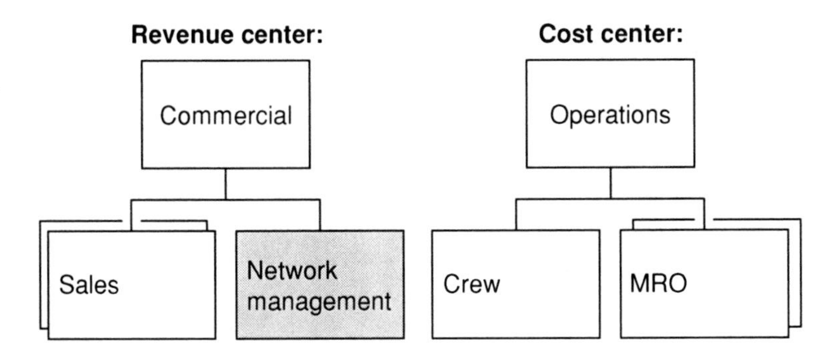

Fig. 75 Network management as part of a broader commercial organizational unit

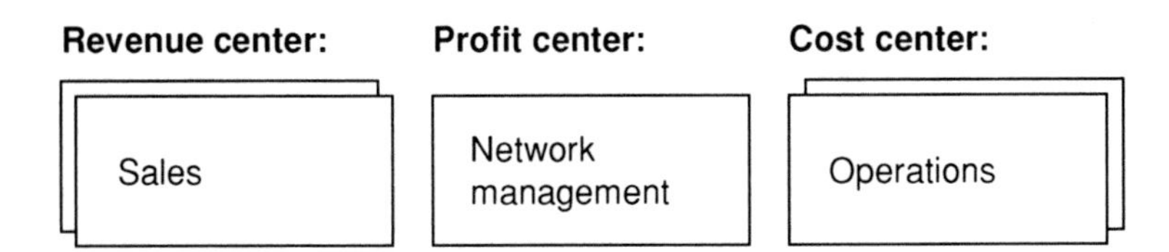

Fig. 76 Network management as the profit center between sales and operations

6.5.1 Structuring by Time Horizon

The rationale for this model is that longer, medium, and short-term network planning have different objectives and require different methodologies. Typically, longer-term or "strategic" network planning focuses on capacity developments and fundamental network design (number and role of hubs). Medium planning departments are frequently called "next season planning" and focus on destinations, frequencies, time-of-day patterns, allocating types of aircraft to specific flights, planning aircraft and crew rotations, and publishing finalized schedules. Short term planning, or "current season," focuses on the short-term alignment of capacities and timings in order to accommodate short-term market variations. This is the typical model for most US and European network carriers.

6.5.2 Structuring by Hub

Some airlines have introduced responsibilities by hub as the prime organizational principle (see Fig. 77). This scheme attempts to emphasize entrepreneurship and the reduction of complexity costs over the full exploitation of synergies. LH is a prominent proponent of this organizational scheme, but has complemented its decentralized hub management responsibilities with a central network strategy and capacity allocation function utilizing a reporting line independent of the decentralized hub managers. This hybrid model offers LH advantages when acquiring airlines in Europe, as the existing network planning departments of the acquired airlines can easily be integrated as additional hub management departments. In addition, each hub management organization can be internally organized by time horizon.

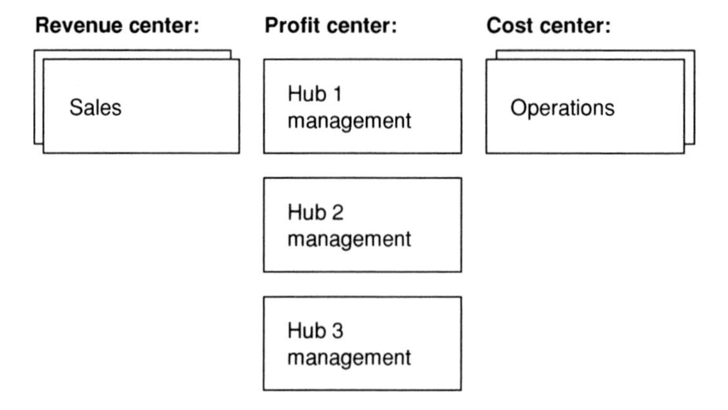

Fig. 77 Hub management as the organizational center of gravity

6.5.3 Structuring by Region

This is the classic organizational scheme of airlines lacking significant transfer volumes. Most major US airlines switched from this organizational model to time horizon based structures when introducing hub-and-spokesystems. In Europe, the connectivity-driven airlines such as LH or AF/KL did the same in the 1990s; this model is frequently found in Asian markets as well. The regional accountability of network planners in this organizational setup would not permit proper planning of transfer traffic, as such traffic often flies through the responsibilities of more than one planner and results in conflicts and suboptimal results.

6.6 How to Make Money Through Networks: The Art of Network Controlling

How profitable is a network? How profitable is a particular flight? How valuable is a specific transfer connection? While the costs and revenues attributable to such

Table 8 Airline variable costs (the comments in the column "Network layer" refer to discussions in Chap. 10)

Contribution margin level	Related cost types	Questions to be asked if contribution level is <0	Network layer
Passenger variable costs	All variable costs related to serving passengers: • Sales commissions • Catering costs • Entertainment • Passenger handling • Variable fuel • Variable crew	Should the number of passengers be reduced (in favor of higher-yield passengers)?	Capacity and pricing layer
Flight variable costs	All variable costs related to flying the aircraft: • Fixed fuel • Airport charges (per movement) • Navigation charges • Variable crew • Variable maintenance	Should the frequency of services be reduced or the particular route be terminated?	Temporal production layer
Aircraft ownership variable costs	All variable costs related to owning aircraft: • Fixed maintenance and engineering • Lease rates and/or depreciation and interest • Fixed crew • Insurance	Should the fleet be reduced?	Spatial layer and demand layer

cost sites must be allocated, how can we allocate the total revenue paid for a connecting flight covering three consecutive legs to these constituting legs? How can we key indirect costs, such as aircraft leasing rates, to individual flights or routes?

When deciding which markets are profitable and how expansion opportunities compare financially, the respective revenues and related variable costs—and resulting contribution margin levels—are the prime focus. While airlines apply different definitions in detail (Morell 2007; Vasigh et al. 2008), they typically differentiate among three types of variable costs and contribution margin levels. In Table 8, we summarize how cost types apply to contribution levels, and which contribution level is needed to answer certain questions.

6.7 A Potentially Dangerous Concept: Route Profitability

In a purely point-to-point world, network controlling is straightforward: Revenues are clearly attributable to individual routes. The same applies for most variable cost components, like variable passenger and flight costs. Policies on how to allocate variable aircraft ownership costs vary from airline to airline. One rule of thumb for keying such indirect variable costs is to distribute fleet or sub-fleet specific indirect costs to routes in proportion to the block time absorbed by this route, out of the total block time of the respective fleet or sub-fleet.

For most planning and monitoring, consideration of passenger and flight variable costs is sufficient. This especially holds true for shorter-term decision making when all fleet-related costs are committed and, from a practical point of view, are effectively fixed. The question of how to allocate these costs to individual components of the overall network does not change them. When planning the expansion of the fleet, however, the respective impact on the network must be understood, including the extra fleet and crew costs.

In cases where individual aircraft carry different lease rates or one aircraft is leased and another is fully depreciated, should such cost differentials be reflected when allocating costs to routes? Unless a particular aircraft is constantly designated to a particular route, they should not: Such costs are driven by operating a fleet of aircraft, not by serving a particular route. The overall costs of ownership remain the same regardless of which aircraft is assigned to which route.

The marginal costs of an underutilized fleet are frequently misunderstood. Some airline managers sometimes calculate the profitability of a route on marginal rather than full costs. Their odd reasoning is that since the capacity is available anyway, putting an otherwise inactive aircraft into service on a certain route would involve little or no extra cost. As a result, a network may be in red ink even though all its individual routes show positive marginal costs. Consequently, marginal cost calculations in route profitability should be considered highly toxic and should be avoided.

6.8 How to Allocate Cost and Revenue for Connecting Flights

6.8.1 Route Profitability and Its Limitations

Route profitability seeks to allocate all relevant revenues and costs to each route served within the respective network. In this framework, each route is considered a profit center. So long as there is little or no connecting traffic within such a network, route profitability calculations are the simplest and most appropriate method for controlling networks. However, this is not the case if the network transports a significant portion of transfer traffic. The passenger pays one amount of revenue for the entire itinerary, and this amount typically is significantly below the combined price of tickets if purchased individually. Assuming a ticket price of 100 euros for a connecting flight consisting of an initial short-haul feeder and a long-haul flight, how can we allocate this revenue of 100 euros to both segments of the passenger's itinerary? How can we allocate the revenue in proportion to variable costs? Or to distance? Or to the first segment, as the segment fully depends on the initial flight to generate the revenue? Or vice versa? In the framework of networks serving high volumes of transferring traffic, individual routes may not be regarded as profit centers. In those cases, the profitability of each route depends on the networked behind and beyond O&D markets. In networks carrying high shares of transfer traffic, route profitability analyses may serve solely as a quick first indicator.

6.8.2 Network Value Contribution

In connectivity-driven networks, we must rephrase the route profitability question from "What is the profitability of a particular route?" to "What is the value a particular route contributes to the entire network, including all behind and beyond traffic?" Simply put: "What are the negative opportunity costs of not serving this route?"

A proven approach to network planning and monitoring focuses on the network value contribution (NVC) of individual routes and works as follows:

- From the standard route profitability figures, all keyed revenues are subtracted.
- Next, planned or empirical revenues for a connecting itinerary are fully (so, double or even triple) allocated to each contributing route.
- Allocated revenues are added up with the genuine point-to-point revenue generated for that route.

This way, each route carries its full opportunity costs: All revenue allocated would be missing if this route were not served (before spill-and-recapture). As a result, the network planner or controller obtains a sorted list with high-value routes at the top, and lowest-value routes at the bottom (or vice versa). Figures will depend on the contribution level of the costs allocated. Great care must be given when interpreting NVC figures: The total of all NVCs in a network by far exceeds the total revenue generated. Therefore, the total NVC is a meaningless number. NVC only makes sense as a powerful indicator of the revenue consequences for the entire network *if not serving a particular route*. A route where the respective NVC exceeds the allocated variable costs provides a positive value, but a route where variable costs exceed the fully allocated NVC should be omitted.

Chapter 7
Competition, Cooperation, Co-opetition, and Antitrust: Balancing Economies of Scale and Competitive Market Structures

Abstract Overlapping networks are the source of competition among airlines, as well as the key driver of synergies in potential alliance expansion or mergers. Network overlap also is increasingly subject to legal battles with antitrust authorities. As demonstrated in the recent merger, synergy, and ATI cases, network overlap has many facets. For that reason, this chapter reviews the various categories of network overlap and assesses the respective competitive and synergetic consequences. We weigh the options of sharing and splitting costs as well as revenues in airline alliances, along with the limitations of such benefit-sharing formulas.

Network overlap is essential to airline competition. Passenger choice only exists if airlines compete for the same origins, destinations, O&Ds, or slots. Understanding network overlap is of interest not only to passengers and airline planners, but also to antitrust authorities and regulators. Network overlap may be defined in the following dimensions:

- Overlapping destinations
- Overlapping (nonstop) routes
- Overlapping transfer itineraries
- Overlapping transfer O&Ds

The more a regional airline market is characterized by transfer traffic (most of the US, Canadian, and European markets), the less relevant are the first two dimensions of overlap, and the more important are the latter two. For example, two strictly route-focused LCCs may serve the same airport but still show zero overlap on route level. Destination overlap is an insufficient indicator of network overlap, since, from the passenger's point of view, the minimum level of competition must take place at the route level. Overlap on a destination level is relevant in assessing dominance on corporate accounts, but not on any segment of the network. For hubbed networks, overlap of itineraries or O&Ds is of much greater importance than overlap of destinations. Network overlap among US majors is usually higher than for European networks. This is mainly due to the enormous geographic

P. Goedeking, *Networks in Aviation*, DOI: 10.1007/978-3-642-13764-8_7,

dimension of the US domestic market, with competing airlines all striving to expand and maximize market share. The resulting network overlap clearly contributed to the bleeding price wars of the past, and has triggered the still ongoing consolidation. Many airline networks are historically (and, in some notable instances, currently are) rooted in highly regulated and protected national networks. Competition among those networks took place on "neighborhood" traffic markets. Former "national carriers" in Europe were slowly entering "foreign national markets" within deregulated Europe. Only the LCCs took advantage of the opportunity from the beginning of deregulation to fill many emerging cross-border or national vacuums.

Another highly instructive perspective on network overlap is the analysis of which portion of a particular network is exposed to which level of competition. Apart from the perspective on overlap in terms of an entire airline network, this analysis becomes particularly instructive if applied to the key hubs of given airlines. Network overlap also plays a key role in designing and managing commuter feeder networks. Most US majors take advantage of a variety of commuter airlines to efficiently feed and de-feed their hubs. They overlap with each other to some extent in order to maintain a minimum level of cost competition among these suppliers.

7.1 Synergies Among Networks: The Key Driver of Consolidation

When an airline joins an alliance, or when airlines decide to merge, they study the potential synergies in advance to justify the move to their stakeholders, the capital markets—and often to themselves. The synergy amounts that are eventually communicated must be accurate and at the same time serve their communicative purpose. Synergy amounts may deeply impact the financial design of the merger or takeover. For instance, consider the question of who "owns" the synergies in a merger transaction: Is it the buyer or seller? In either case, any premium paid must be fully financed by synergies. If synergies are expected to materialize early, the associated risk of the acquisition is low, thus the price premium should be high. If synergies are more likely to come into effect late, the price premium must be lower, reflecting their higher risk and lower NPV. If the price premium is high and synergies come late, the deal has failed the moment the contract is signed (Sirower 1997).

In all airline synergy analyses, networks play the central role. During recent mergers, some news commentators viewed the *lack* of network overlap as an indicator of synergetic complementarity, while others believed the degree of network overlap was driving synergies. For some, overlap is a source of cost synergies; for others, it is a source of revenue synergies. Therefore, some clarification is needed.

Moreover, much dispute has focused on how to reliably assess synergies, be it a bottom-up methodology (the airliner's preference) or a top-down (the investment banker's choice) approach. In particular, this applies to network synergies:

- Bottom-up methodologies analyze feasible synergies in all relevant functions and business segments. Airlines usually apply these approaches, as they provide the most detailed view of where synergies might originate and in what order of magnitude.
- Top-down approaches use benchmarks and econometric models to assess synergies.

Both approaches have their strengths and their weaknesses. Bottom-up approaches can be more detailed, but are much more vulnerable to biased opinion—the synergies generally will not be realized in the expected areas. On the other hand, top-down methodologies may be more efficient, but cannot be as detailed. Experience shows, however, that both schemes convert to similar overall conclusions.

Network-related synergies are important because of their amount and their relative contribution to the airlines involved. They have a deep impact on the governance of the respective alliance or merger.

The statement that two networks would "nicely complement each other" is sometimes overestimated, as the most significant and quickest synergies do not come from an expansion of the network scope. The high potential—and the truly high-hanging fruits—will be found in leveraging network overlap and market clout, through:

- The application of advanced revenue management techniques on overlapping markets
- Capacity reductions or schedule alignments on overlapping monopolized markets
- Joint sales management of corporate accounts on overlapping markets

Networks that do not overlap will not be able to offer any significant cost or revenue synergies. In contrast, if two networks fully overlap, one is superfluous. Hence, careful analysis of the amount and structure of network overlap is essential to understanding network synergies.

7.1.1 Network Synergies are Difficult to Monitor

During a merger, the talk at press conferences is all about prospective synergies, while published figures on the success of actual synergies are much vaguer. No outside observer, financial analyst, or industry expert can deduce from balance sheet or profit-and-loss data the time, type, and scope of the synergies that have been cashed in. Conversely, few airlines can accurately assess how much of their cost savings or revenue flows stem from synergies or other sources. When Austrian Airlines switched from Qualiflyer to Star Alliance back in 2000, they hoped to eliminate the bleed due to overcapacity assigned to high-yield traffic between Germany and Austria. That hope may or may not have been satisfied in the years that followed. Meanwhile, however, the low-cost carriers heavily affected yields between Germany and Austria. Hence, it is impossible to assess how much of the synergies promised yesterday are being realized today. In an industry with low barriers to entry and dynamic competition, synergies rarely match the forecasted figures, just as reality rarely resembles the forecasted scenario.

7.1.2 Synergies are Not Always the Top Priority in Mergers

While the capital markets may always factor in all kinds of fancy synergies as real and given, more airline executives are refraining from rigid synergy implementation. The reasons are mainly twofold:

- If a high proportion of cost synergies is due to redundancies, a rigid policy of eliminating jobs to suit capital market expectations may be to the detriment of any viable service culture, and thus counteract the desired synergies. Yet, there may be differences among the United States, Europe, and Asia, where this cultural effect really matters.
- The second reason is governance related. The realization of network-related synergies (schedule and revenue management synchronization) clearly requires a centralized network management organization and respective accountability. However, the concept of decentralized hub managers, as in the case of LH, emphasizes decentralized entrepreneurship, as well as internal competition over maximum exploitation of theoretical synergy opportunities ("management of networks over network management").

7.1.3 Network Synergies: Value for Whom?

From a financial or numerical point of view, the value represented by synergies may be the same for both partners in a merger. Yet, the same numerical synergy figures may mean something different to each partner in terms of strategic impact—and well before any consideration on who is buying and who is selling.

Consider the following two cases. First, let us assume a case where network overlap represents a fraction of a major airline's network, whereas the same overlap equates to the bulk of the smaller partner's network (see Fig. 78). A situation like this is more about the bigger airline swallowing the smaller rather than about the two cooperating.

In the second case, two networks heavily overlap, reaching deeply into the core of both airlines' markets. Experience shows that too much overlap is more likely to result in conflicts than synergies. The symmetric or asymmetric proportions of network overlap are at least as important in understanding the pros and cons of a proposed alliance or merger as is the raw amount of synergies or overlap. This is the area where the potential for a conflict between stakeholders and shareholders is higher, with its implications affecting the long-term success of the proposed merger.

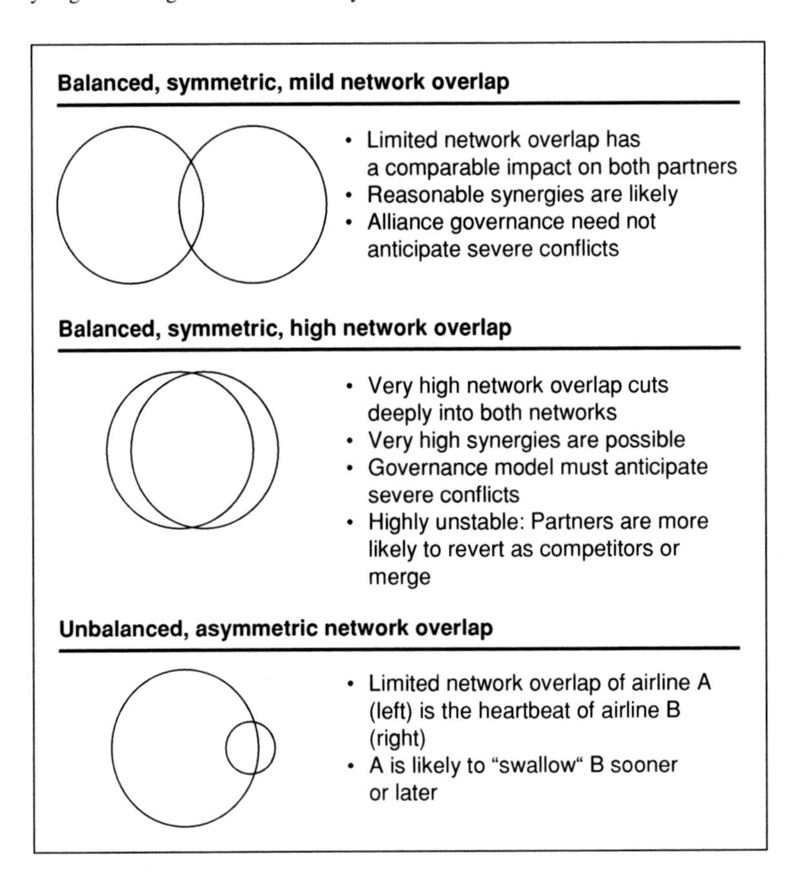

Fig. 78 Value for whom? The total amount of network overlap and the distribution of network overlap between the two partners are equally important

Network synergies may be a source of synergies and competition at the same time: some airlines are fierce competitors on most parts of their respective networks, but still form cooperative codeshare agreements on selected routes. Brandenburger and Nalebuff (1996) formed the term "co-opetition" for this kind of phenomenon. With codeshares common on nearly 14% of all transfer connections worldwide, the aviation industry may serve as an advanced example of co-opetition being commonplace.

7.2 Synergy Sharing Formulas: Simple or Fair, but Never Both

Airlines frequently cooperate, either per route in the format of codesharing or overarching as in global alliances. For cooperative services to selected markets, the cooperating airlines must agree on how to share or split costs and revenues. For the simplest codeshare agreement on a particular nonstop route, mechanisms for "prorating" are well established. To accommodate more complex codesharing on transfer routes, airlines complement the simple prorated agreements with extra agreements. Cost–benefit sharing becomes complex, however, when airlines agree to cooperate on broad parts of their networks. This is the case when airlines seek to cooperate on long-haul markets like the North Atlantic. All formulas on how to split and share costs and revenues inevitably suffer from one weakness: those formulas are either simple or fair, but never both. In Sect. 6.8 in Chap. 6, we discussed the complex process of assessing the network value of a single route that carries significant before and behind transfer traffic. If it is difficult to assess within the closed framework of an airline management accounting system, how much more difficult is it to plan and monitor in between airlines, each with its own balance sheet and shareholders? Far-reaching cooperation involves large amounts of money. Consequently, cost and revenue-sharing formulas must be approved by high-ranking supervisory or shareholder committees. Formulas that are likely to find the approval of network experts are unlikely to be easily understood by shareholder committees. On the other hand, formulas that are easily understood by shareholder committees are unlikely to find the blessing of network experts. Moreover, if fairness requires complex formulas, the effort and related costs to monitor such complex formulas, which must be repeated on all sides of cooperating partner airlines, may counteract synergies. Extensive cooperation and full exploitation of synergies require one single cash box—all else remains suboptimum.

7.3 Antitrust Immunity: Exempt from the Rules or Rules of Exemption?

Network overlap has increasingly become an issue in antitrust (ATI) decisions on airline mergers. The challenge for regulatory authorities and airlines in ATI cases is the appropriate definition of "What is a market?" In the past, antitrust authorities have often taken a simple "overlap by destination" or "overlap by route" view to judge on the competitive impact of merging transfer and traffic-driven networks. This has led to approving inappropriate mergers and to denying (or curtailing by destructive constraints) mergers that should have been approved if they had been measured by the right level of competition. To give a few examples: if a selected long-haul route carries 80% transfer traffic, the competitive position of this route may only be assessed by comparing all itineraries with any behind origins or beyond destinations using this selected segment (see also Fig. 5). If the competition is about an origin market (see Sect. 1.3 in Chap. 1), then the question of slot dominance is appropriate. However, if O&Ds could just as easily connect via other itineraries, slot dominance at one selected point of transfer does not necessarily distort competition. In the recent past, applicants, opponents, and regulatory authorities alike significantly contributed to the erosion of correct

Table 9 Objects of competition and related indicators of potential competitive dominance

Object of competition	Indicator of competitive dominance for a particular market participant
Origin or source market (see Sect. 1.3 in Chap. 1)	Share of outbound movements, ASK, seat capacity, MIDT or BSP bookings, sales with local corporate accounts, gate, apron, or counter positions
Destination market (see Sect. 1.4 in Chap. 1)	Share of inbound movements, ASK, seat capacity, gate, apron, or counter positions
Point of transfer (hub)	Number of overlapping transfer itineraries at alternative transfer hubs (absolute and weighted by connecting seats)
Routes (see Sect. 1.5 in Chap. 1)	Share of movements or seat capacity
Transfer O&Ds (see Sect. 1.5 in Chap. 1)	Number and value of overlapping transfer itineraries via alternative transfer hubs

market definitions by using various definitions in court, as deemed necessary (understandably but shortsighted), to win the case at hand. It is important for regulatory authorities to consistently apply more apt market definitions when they take the stand, even though these definitions may be more complex and difficult to assess from a legal point of view. Table 9 characterizes the various markets and related types of competition, and identifies the appropriate performance indicators for judging the potential need and the measures for protecting level playing fields.

Chapter 8
Multi-Hub Networks: Masterpieces or Nightmares of Complexity?

Abstract Soon after deregulation, networks with multiple major transfer hubs emerged in the United States, and European and Asian airlines quickly followed. In the recent past, many airline mergers accelerated this trend. Networks with multiple hubs cover more regional space and handle more transfer traffic. On the other hand, multi-hub networks exhibit significantly more overlap with competing multi-hub networks, spurring competition but also accelerating yield erosion. In addition to expanded regional scope and scale, multi-hub networks offer passengers a variety of itineraries for outbound and homebound transfer journeys, all within the same tariff structure. Multi-hub networks often offer multiple itineraries on the same O&D at similar times of day, thereby creating potential internal competition. This chapter examines the strengths and weaknesses of multi-hub structures, and introduces the tactics for assessing and improving synchrony within such complex networks.

So far, we have assumed a single-hub network with related spokes to feed and defeed the respective hub. However, networks of large regional scope and high-density services can strongly benefit from a multitude of transfer hubs. Single-hub systems reach a naturally sized ceiling when too many important transfer connections require excessive detours. To give an example, a high-performance hub on the US East Coast is of little value to serving North–South traffic along the US West Coast. Networks expanding into additional markets must eventually establish new hubs to competitively serve these new regions. When acquiring another airline, the acquiring airline often inherits a single-hub or multi-hub network that needs to be integrated.

However, it can be shortsighted to conclude that nearby hubs partially serving the same O&Ds are redundant and that one of the overlapping hubs should be omitted. Multi-hub airlines quickly discovered that the combination of even significant "internal" overlap among various hubs with state-of-the-art revenue management techniques can yield enormous benefits. As a result, airlines now

P. Goedeking, *Networks in Aviation*, DOI: 10.1007/978-3-642-13764-8_8,

emphasize internal overlap and actively develop nearby hubs (AF/KL hubs at CDG and AMS or LH hubs in FRA, MUC, and ZRH). Widespread in the United States for many years, multi-hub networks quickly emerged in Europe but have not yet developed in Asia.

8.1 Prototypes of Multi-Hub Strategies

Multi-hub systems emerged in the United States in response to deregulation in 1978. Up to the early 1990s, when deregulation started in Europe, "domestic" referred to a plethora of small or midsized states, each with one "hub" (usually the capital) served by the respective "national" or "flag" carrier. As a result, Europe was covered by a multitude of scattered hubs, but those hubs were isolated from each other, not part of a synchronized network. Today, this picture has fundamentally changed and continues to evolve. Meanwhile, AF and KL have integrated and synchronized their prime hubs in AMS and CDG. LH has developed MUC in addition to FRA; has acquired LX with its ZRH hub, SN in BRU, and OS in VIE; and is about to develop MXP into its northern Italian hub.

Airlines follow different strategic objectives in developing their multi-hub systems. Two distinct multi-hub strategies prevail:

- *Scale strategies* aim to optimize transfer traffic volume and spatial coverage. Scale-driven multi-hub strategies typically apply to large single markets with centers far apart from each other. In theory, overall connectivity in a multi-hub network is best served with as few but big as possible hubs. Since scale is more about unit cost advantages, high overlap between network internal hubs is considered disadvantageous. The traditional objective of US-based multi-hub networks is to provide optimum regional coverage. While high overlap of transfer traffic between competing hubs drives the fight for market share, it can add pressure on yields and raise questions about network restructuring in mergers.
- *Scope strategies* emphasize yield over volume. To do so, scope strategy-driven airlines focus more on the dominance of high-yield source or origin markets rather than on the dominance of O&Ds. Europe offers relatively more high-yield markets than the United States, permitting—and to some extent requiring—more hubs nearby. Therefore, scale strategies appear highly adaptive to the US environment but less adaptive to the European landscape, while the opposite applies to scope strategies. US majors typically follow scale; European majors typically scope strategies.

After their merger, AF/KL moved quickly to fully integrate their tariff structures. That way, they could offer passengers on their morning outbound flight the possibility to connect via CDG and to connect homebound in the evening via AMS, all within the same tariff structure and system. This mechanism is known as "fare combinability." It requires that on each O&D both airlines offer the same booking classes, the same tariffs per booking class, and the same applicable restrictions per booking class, and a fully transparent booking and inventory system. As a result, passengers perceive a tariff system provided by one single airline. The costs of implementing fare combinability in terms of required IT investments are usually high—and so are the benefits if applied to strongly overlapping markets.

8.1.1 Scale, Scope, and Network Overlap in Multi-Hub Networks

Particularly in multi-hub networks pursuing scope strategies, significant overlap between the participating hubs can be found. Serving 50% of O&Ds via more than one hub in a multi-hub network is not unusual. If the strategic focus is on scope, or on adding high-yield POS markets to the network, significant overlap may occur. Such internal overlap may result in:

- Redundant capacity
- Thorough fare combinability

When DL acquired competitor NW, formerly competitive overlap between hubs became internal overlap: NW's former hub at DTW significantly overlaps with DL's hub at CVG, and NW's former hub at MEM strongly overlaps with DL's hub at ATL. Thus, capacity at both hubs is under scrutiny. For CVG, a decrease in destinations from 140 four years ago to 70 has been announced (Keypost 2009). After their merger, AF and KL emphasized growth over costs rather than reducing capacity. They have introduced thorough synchronization of O&Ds (see Sect. 8.3.1) and fare combinability.

8.2 Self-Amplification of Growth

Figure 79 shows that large hubs in Europe enjoy a strongly overproportionate share of transfer traffic compared to smaller airports. All else being equal, this comes at the expense of smaller airports. Obviously, a threshold of critical mass, also called a "hub effect" (Goedeking et al. 2009), provides a further boost in growth for the airports shown. Below this threshold, the very same effect works against the airports that are too small.

What is the driving force behind the hub effect? As shown in Sect. 3.1 in Chap. 3, the number of hits follows the square of the number of underlying flight movements at a given hub. Clearly, passengers prefer connecting via hubs that offer many connecting opportunities. Passenger preference is driven not only by the quality of an individual connection, but also by the quality of the point of transfer. This finding has fundamental consequences for multi-hub strategies: Given the overproportionate effect, one large hub attracts significantly more transfer passengers than two hubs of half the size. Consequently, an airline has a

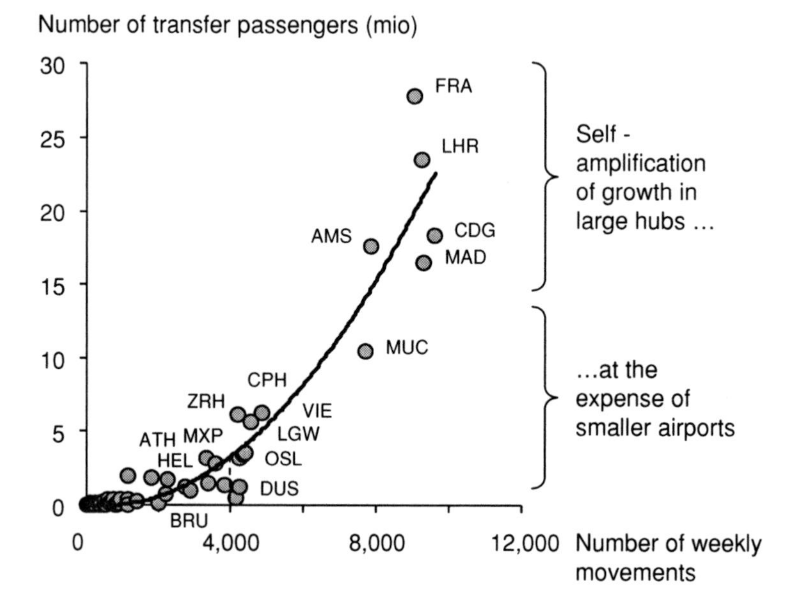

Fig. 79 Self-amplification of growth for European airports (data for summer 2007)

vested interest in operating as few, but as large as possible hubs. Multi-hubbing can only be justified by spatial expansion, by the need to improve market clout on a smaller hub characterized as a high-yield POS, or by the fact that capacity constraints no longer permit growth within the given hub's infrastructure. The latter example also shows why a municipal airport system cannot be as transfer efficient as a single hub: infrastructure fragmentation severely weakens the transfer efficiency of a network.

8.3 Synchronizing O&Ds in Multi-Hub Networks

In a network of multiple hubs, synchronizing the time-of-day patterns of various itineraries on the same O&D can contribute significantly to the quality of service offering and competitiveness. If, for instance, a US coast-to-coast O&D is offered via two different transfer hubs but at comparable overall departure and arrival times, then the two itineraries compete against each other for the same passenger segment. The respective airline competes against itself by virtue of redundant itineraries. Sometimes, such patterns of poor synchrony can also be found when analyzing the combined schedules of alliance partners. In a well-synchronized multi-hub network, various itineraries complement each other. One itinerary, for instance, is served at each even full hour at the origin, while the other itinerary is served at every odd full hour (Fig. 80).

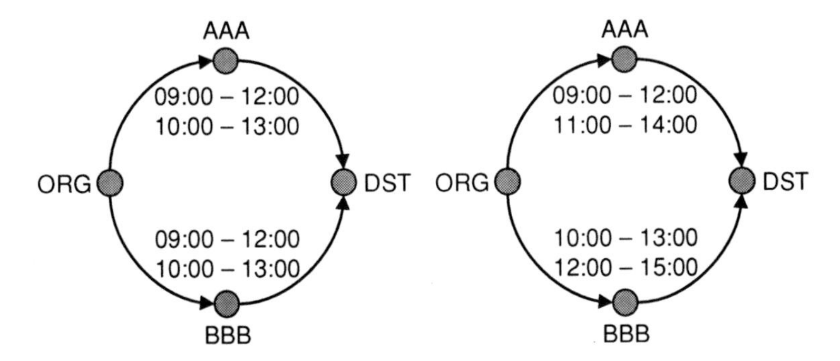

Fig. 80 Synchronization in a multi-hub system. On the *left*, the O&D ORG-DST is served via AAA and BBB. Connections via AAA and BBB all depart from ORG and arrive at DST at the same time, cannibalizing each other. In the schedule shown on the *right*, connections via AAA and BBB, respectively, alternate, thus complementing rather than cannibalizing each other

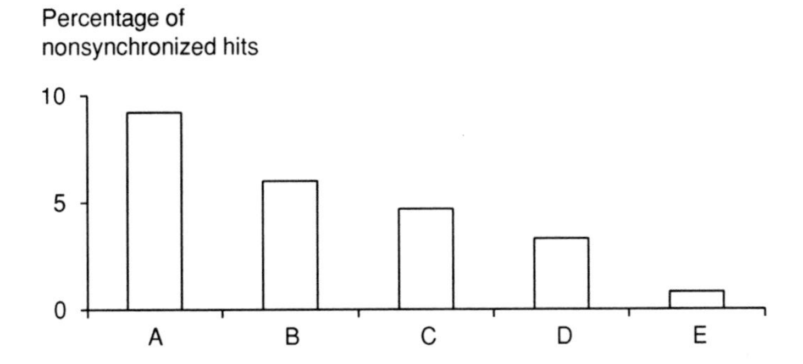

Fig. 81 Level of multi-hub synchrony for a selection of major US airlines

8.3.1 How to Identify Insufficient Synchrony

A given transfer connection is not synchronized if another nonstop or transfer connection within the network of the same airline (or alliance system) departs at or after 30 min before the STD of the initial segment of the transfer connection at hand, and arrives no later than 30 min after the STA of the final segment of the connection. The buffer of 30 min may be varied depending on various market aspects. In Fig. 81, the portion of nonsynchronized hits out of all hits is shown for a set of major US airlines. The figures boldly underline the importance of properly synchronizing alternative itineraries within a multi-hub network.

Chapter 9
Assessing and Comparing the Strengths and Weaknesses of Aviation Networks

Abstract In this chapter, we introduce the benchmarking of networks and answer the following questions: How well do networks perform in comparison with others? What characteristics can—and cannot—be compared? What quantitative key performance indicators (KPIs) are appropriate? The various categories of benchmark network characteristics are examined and complemented by selected examples of specific KPIs.

By comparing the structural characteristics of networks based on quantitative KPIs, benchmarks serve as a way to uncover relative strengths and weaknesses and, most importantly, to unearth the network strategies of competitors. Because the geographical, political, demographical, regulatory, and competitive environment is different for each airline, the network strategy should be as well. Care must be given, therefore, not to compare apples with oranges and draw inappropriate conclusions. For instance, a TAT of 20 min for an LCC should exert pressure on a network carrier to improve efficiency and productivity. However, network carrier procedures should not be considered inferior if they take 30 min for a comparable type of aircraft. The most important benefit of benchmarking is that it can help profile the strategic intent underlying the apparent network characteristics. Benchmarking also may shed light on the hidden risks of networks, thus providing valuation indicators for an airline operating a particular network.

To draw a complete profile of network characteristics, strengths, and weaknesses, we must consider six essential KPI categories:

- Demand volume and structure
- Production, productivity, and capacity
- Connectivity
- Geographical scope
- Risk exposure
- Trends of all the above

P. Goedeking, *Networks in Aviation*, DOI: 10.1007/978-3-642-13764-8_9,

Most KPIs are important systemwide, by hub (airline perspective), hubwide, or by airline (airport perspective). The same applies for historic trends. KPIs may slightly differ between airlines and airports.

9.1 How to Benchmark Demand Volume

An airline or airport may serve many city pairs, operate many flights and destinations, and exhibit the highest scores of connectivity. All this is worthless, however, if demand is insufficient to fill the respective aircraft. Sufficient volume (and structure by yield) of passenger demand is the heartbeat and backbone of any network. Therefore, it is crucial to understand how "thin" or "fat" the O&Ds of a given network are. Clearly, the higher the share of medium or fat O&Ds, the better, as a network suffering from prevailing thin and fragmented O&Ds will incur relatively high operating costs.

However, demand volume data are also vital to the correct weighting of hits, network overlap, and other KPIs. Demand volume in itself is an excellent indicator of the geographic location quality of a hub by cumulating the total O&D demand of all O&Ds served by a particular hub, airline, or airport. When compared to the market-specific quality of service data (QSI, see Sect. 2.3.4 in Chap. 3), passenger demand volume data are indispensable in identifying overserved or underserved markets and growth opportunities.

Figure 82 compares the O&D portfolio of two major European airlines at their respective main hubs, differentiated by nonstop and transfer O&Ds. Airline A (on

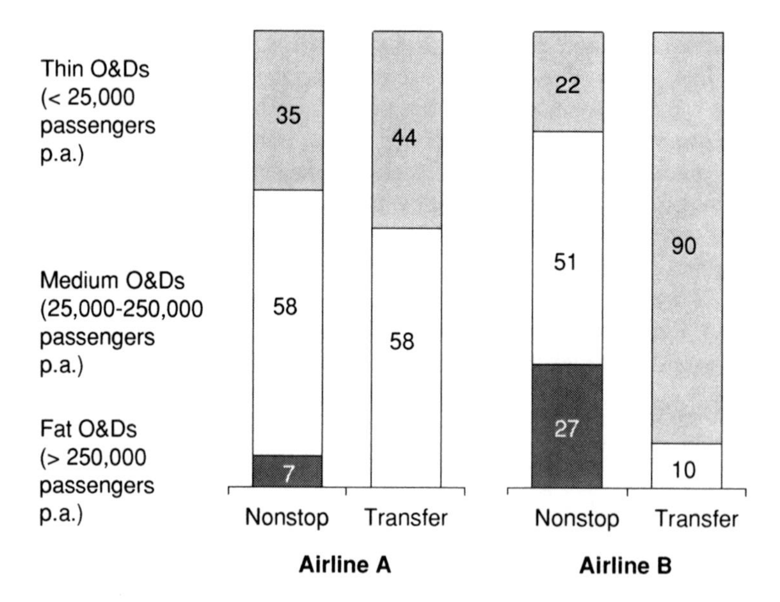

Fig. 82 Benchmarking O&Ds of two large European airlines by demand categories (percentages)

the left) represents a network with a limited number of fat or medium O&D markets; it is compensating for this weakness with a relatively high share of medium transfer O&Ds. In contrast, Airline B (on the right) can build upon a high share of medium and fat nonstop O&Ds, but lacks a significant amount of medium transfer O&Ds. These figures may be either the cause or the result of the underlying network strategy. They clearly show, however, that Airline A depends on transfer traffic, while Airline B depends on the strong demand for nonstop traffic.

9.2 How to Benchmark Productivity and Capacity

The utilization of costly resources like crew and aircraft has a direct impact on the bottom line. The benchmarking of utilization figures serves to challenge current operational procedures.

The term "capacity" has different meanings depending on whether it is used in context with airports or airlines. For airlines, capacity refers to the number of flights, seats, or gates; for airport managers, the same term refers to runway capacity. Therefore, the term "production volume" is used here to refer to the number of inbound or outbound movements, routes or destinations offered in a given network, seat capacity, available seat kilometers (ASK), spiked wave patterns (measured, for instance, as the standard deviation of the envelope of inbound and/or outbound movements at a given hub), or aircraft utilization. Particularly for airports, the spiked movements at the respective airport are a good indicator of how well an airport's key assets, the runway and the terminal, are utilized over the course of the day.

Another instructive indicator of productivity is schedule consistency. Schedule consistency measures the ratio of routes served Monday through Friday with consistent STD, STA, aircraft type or family, and flight number over all routes. A low score of consistency is usually an indicator of extensive short term fleet assignment. A high score of consistency frequently is the result of advanced pricing mechanisms.

9.3 Why Benchmarking Connectivity is Vital to Uncovering Network Strategies

As Fig. 82 shows, some airline strategies emphasize local traffic, while others focus on transfer traffic. The more an airline or hub depends on transfer traffic, the more important high performance is to connectivity. Understanding the connectivity of a particular airline or airport in relation to its competitors is invaluable in optimizing the given network to attract as much valuable transfer traffic as possible. Losing ground in terms of connectivity on key transfer O&Ds will quickly translate to market share loss and will likely spiral downward to cash loss.

Connectivity-driven KPIs may become complex, depending on the depth of required analysis and competitive comparison. Benchmarking the total number of hits (the hub perspective) or normalizing hits to the number of inbound flights (the passenger's perspective) or to the number of expected hits (see Sect. 4.2 in Chap. 4) represent straightforward connectivity KPIs. Benchmarking connection-timing profiles (Fig. 19) to assess how well connecting flights are synchronized to complement each other rather than form redundancies is demanding.

The most fundamental indicator of connectivity is the absolute number of online hits by an airline at a given airport or systemwide. Care must be applied when including code-shared flights in connectivity benchmarks. Connectivity is mainly driven by online connections; while interline or code-shared connections can amplify basic connectivity, they rarely can determine bank design.

Aside from the absolute number of hits, airlines need to understand how many de-feed flights are available per inbound flight on average, or vice versa. Thus, the KPI "hits/number of inbound" is an easy-to-apply yet powerful indicator of connectivity in the context of de-feed capability. Also, passengers are more interested in the number of hits offered upon arrival at a transfer hub rather than in the total number of hits offered at a given hub. Again, "hits/number of inbound" would provide the appropriate KPI.

Both the absolute number of hits and the hits/inbound, to a lesser extent, are biased by the number of underlying flight movements. (See Sect. 4.2 in Chap. 4 for the rationale behind this.) Two KPIs offer a connectivity indicator that is free of such bias: "hits/inb*out" or "hits/expected hits." In the first case, the number of hits is normalized against the theoretical maximum possible number of hits; in the latter case, the number of hits is normalized against the number of hits expected for inbound and outbound flights randomly distributed over the course of a day.

So far, all hits have been valued equally. In reality, however, some hits are more valuable than others. When designing or evaluating schedules, the scheduler needs indicators to emphasize highly valuable hits and de-emphasize connections of limited value. Hits must be weighted according to the desired aspects of value or importance.

Following the above definition of hits, a hit between two 50-seaters and a hit between an A321 and an A380 would count the same. To correct this weakness, we must weight the hits.

Danesi (2006), Bootsma (1997), Burghouwt and de Wit (2005), and Veldhuis (1997) have proposed QSI methodologies to weight connections. Burghouwt and de Wit (2005) have developed the "weighted indirect connection number," whereby they take two ratios (connection detour over greater circle distance, and transfer time over maximum connecting time), and then weight each ratio with respective factors. The total connectivity of a given hub is the total of all "weighted indirect connection numbers" of all connections. Danesi (2006) has introduced weighting factors to complement the original "connectivity ratio" developed by Doganis and Dennis. These approaches all fall short of the logit or neural network methodologies for assessing the appeal that a particular hit or total hub schedule may have in attracting transfer passengers. The complexity of weighted connections based on

QSIs is not much different from logit. Given this methodological weakness on the one hand, and the conceptual complexity on the other hand, the more powerful logit or neural network methodologies are preferable.

9.4 How Big is a Network? The Importance of Benchmarking Geographical Scope

The size of a network matters: the bigger a given network, the more likely are scale benefits (unit cost advantages or improved market clout). Size may stem from high-frequency services to a limited number of destinations, many destinations served at moderate frequencies, relatively large types of aircraft, or emphasis on medium-haul and long-haul flights.

The regional scope of an airline network may be assessed, for instance, by counting how many destinations, cities, or IATA regions are served by how many flights per week. It is important to use data of identical quality when benchmarking such data. Some airports, for instance, use data stemming from the respective national slot coordinator. They offer 100% precision for the respective country but rarely for other countries, thus prohibiting cross-border benchmarks. Others rely on electronic schedule data, constrained by the lack of charter flight data. Clearly, data from both sources are not comparable, and if mixed would severely skew the results.

9.5 What Risks are Associated with a Network, Compared to Others?

For too long, the various kinds of risks inherent to specific network strategies and structures have been ignored or underestimated. With more airlines and airports exposed to the pressures of capital markets, the management of network-related risks is gaining ground. Risk may be reflected in the valuation of airlines, airports, and their respective stock, but also may play an important role in better balancing risk and opportunity in a specific network's growth plans and resource allocation. With airports increasingly asking for insurance coverage of flight cancellations, insurance companies are keeping a careful eye on identifying and assessing network-related risks.

Network-related risks may stem from:

- Significant network overlap with competitors
- Political, military, financial, or social risks of countries or destinations served
- Structural operation weaknesses, such as tight connections causing a high risk of delays
- Regulatory exposure due to high levels of CO_2, NO_x, noise, or other emissions

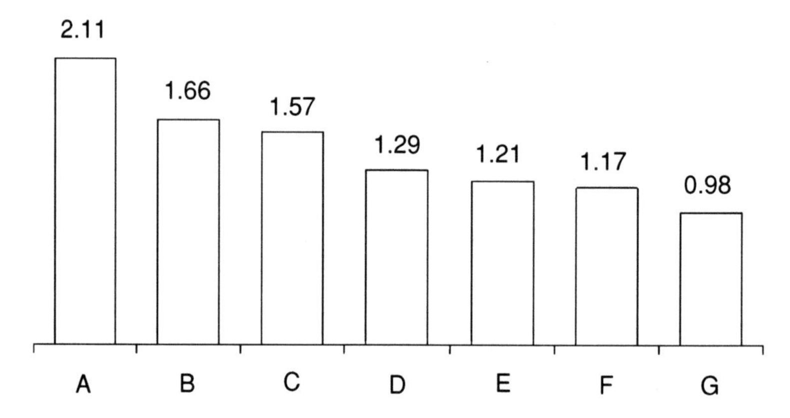

Fig. 83 Benchmarking the average default probability (in %) of demand in the overall network of selected European airlines

To determine network overlap risks, detailed analyses of network overlap are needed. For country-specific risks, the share of seat capacity of a given airline as exposed to countries and their respective risk category must be computed. Country-specific risk data may be obtained from governmental organizations like the International Monetary Fund (IMF), the World Economic Forum (Schwab 2009), or commercial vendors. Sound indicators of country-specific risks of demand volatility are the so-called "credit default swaps," or CDSs [relating to either credit risks in the private sector or governmental bonds (sovereign CDSs)]. CDSs measure the interest payable to cover the default risk of bonds or credits. The interest can be converted into default probability (credit default probability, CDP), resulting in a proxy of demand volatility. The seat-capacity weighted average country risk of an airline's network is a powerful indicator of its vulnerability to unexpected and country-specific risks of collapsing demand.

In Fig. 83, we compare the risk of demand default for some European airlines. The spread of such risk among the carriers shown is surprising, with some airlines being exposed to twice the risk of others. Those risk indicators reflect the risk inherent in the portfolio of destinations served by a given airline or airport. Those indicators are as equally important to resource allocation decisions in "risky" airlines as they are to the valuation of airline and airport stocks.

9.6 Some Network Strategies Cannot be Identified by Isolated KPIs—they Require Analysis Over Time

Not only can the evolution of KPIs be tracked over time, but the future development of KPIs also can be anticipated since airlines publish their schedules many months in advance. Such anticipation can serve as a powerful early warning system in the context of competitive analysis. The evolution of selected KPIs can unearth underlying industrial trends, such as the sharp decline in demand and subsequent capacity deployed on most markets worldwide soon after the 2008 recession hit the market, or the slow recovery since late 2009. Careful analysis of trends can also exhibit a change in the strategy of specific competitors.

Chapter 10
An Overarching Concept: The Hierarchical Layers of Aviation Networks

Abstract In this chapter, we introduce the concept of hierarchically layered (or networked) networks as the framework for a more advanced understanding of aviation networks, including the organizational processes of network management and network controlling.

10.1 Networks are Layers of Networks

An airline network is more than just a level plane of nodes and edges as described in Sect. 1.1 in Chap. 1. In reality, an airline network represents the integration of interdependent and hierarchical layers of specific networks. Each layer adds a distinct set of new paths or edges, serving a layer-specific purpose. By doing so, each layer builds upon the paths or edges of the underlying layer.

To give an example: at the lowest level, a set of edges is defined representing passenger demand to connect two cities. Building upon this basic set of demand edges, airline planners first carve out a part of the demanded overall journey that could be served by flight service, assuming the remaining parts of the overall journey will be served by other modes of transportation (car, train). As a result, the original single edge of demand is now represented by a path through three distinct edges: the demand edge is translated into a basic production path. Airline planners then take the "flight segment" and break down this edge into two edges of connecting flights, thus adding a level of operational efficiency. The result is a set of layered networks, each building upon the other. If seen "bottom-up," each "higher" layer of the stack of networks breaks down a lower level of edges into a series of subsequent edges, or paths. If the same stack of layered networks is seen "top-down," a path at a higher network level is aggregated as a single edge at a lower level.

Airline networks are always composed of many distinct layers of networks, each building upon a subordinate one, and also receiving feedback from "higher"

P. Goedeking, *Networks in Aviation*, DOI: 10.1007/978-3-642-13764-8_10,

layers or providing forward information upward. With each layer satisfying the strict definition of a network, an airline network can best be understood as a layered network of networks: (Fig. 84)

Burghouwt (2007) proposed a two-dimensional framework to differentiate the spatial and temporal characteristics of airline networks. While this differentiation represents a major step forward, the layered rather than two-dimensional framework is more conclusive in relation to the structural prototypes, planning procedures, organizational consequences, and economics of airline networks.

Layered network structures also may be found in other areas, including the design of multilayered electronic circuitries, neural networks, or software engineering. However, in all these applications, the various layers follow the same rules or syntax. In neural networks, for instance, the various layers of "neurons" follow exactly the same syntax. The same applies to the design of electronic circuitries. In airline networks, however, each layer is defined by a distinct set of

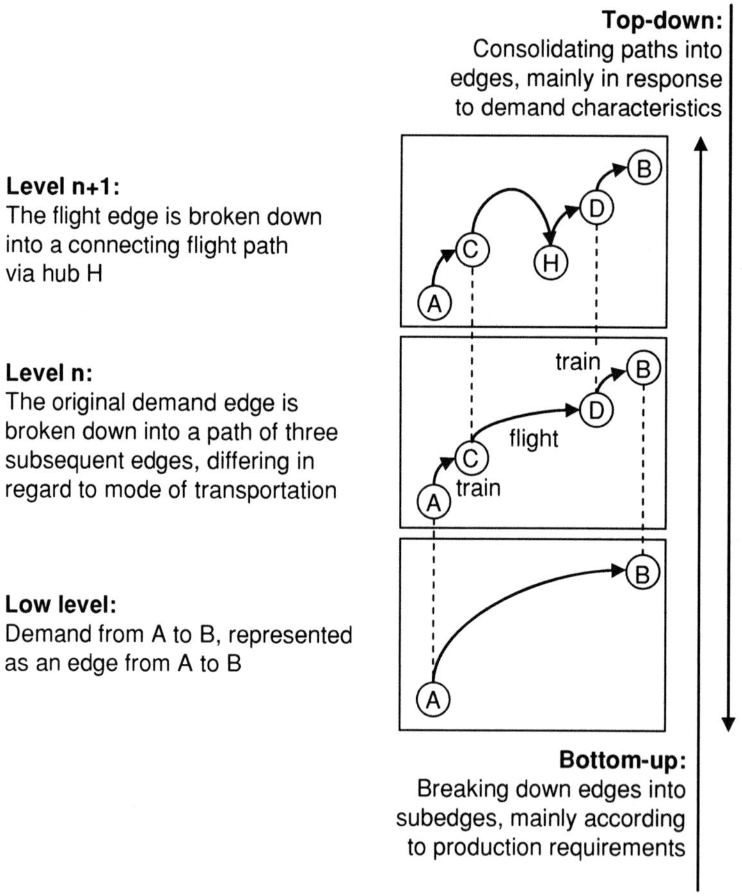

Fig. 84 Paths are edges of lower layers of the same network, and vice versa

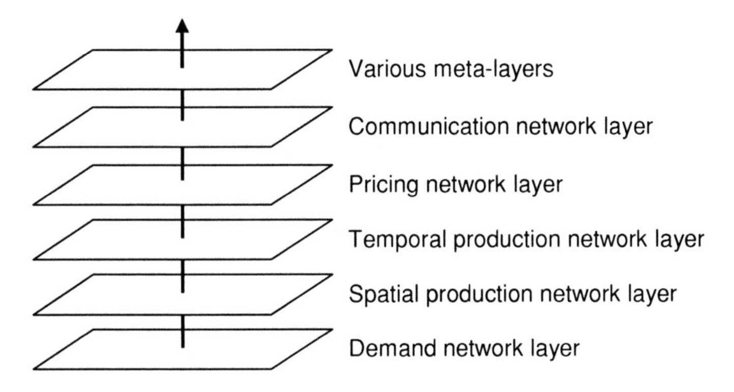

Fig. 85 The various layers of an airline network

edges, and the syntax or rules of interaction are different at each layer: one layer reflects demand; others "speak" operations or "passenger communication." Layered airline networks follow a distinct syntax at each layer. To that extent, the layers of networks—or networked networks—make it tempting to conceptualize airline networks not only as layers of two-dimensional networks, but also as a single three-dimensional network.

In Fig. 85, we examine in more detail the key layers of an airline network.

10.1.1 Demand Structures are the Basis for all Networks in Aviation

At the lowest level, an airline network builds upon the network or networks represented by the passengers or companies demanding to travel. With individual

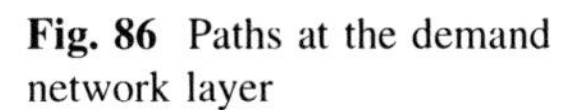

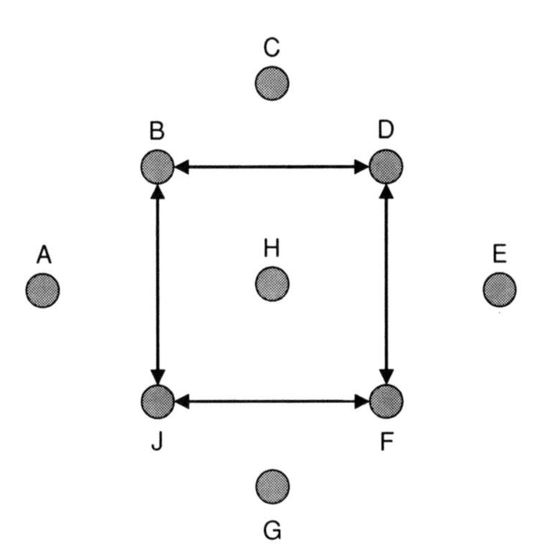

Fig. 86 Paths at the demand network layer

passengers desiring to fly from an origin to a destination, the passenger demand network is composed of a collection of loosely interrelated edges connecting the points of origin with the destinations (nodes) to the demand for flight services (edges). An interesting aspect of the demand layer networks is that they are inherently "distributed" in terms of their topology: demand rarely desires connections (centric network topologies), but nearly always prefers nonstop (Fig. 86).

10.1.2 A Spatial Network is Superimposed on the Demand Network

At the next higher level, airlines must decide on a fundamental production issue. Given the set of isolated destinations (nodes) and edges connecting them at the demand level, airlines must opt for a new set of edges. They may be unable to serve 1:1 the connections demanded by the market, and city pairs may be too isolated or too limited in demand for an aircraft that is too large. So, airlines must find a way to link all key destinations with each other from the production point of view. For instance, they may opt to link some destinations by transfer services rather than nonstop, replacing some nonstop edges with "connecting" paths.

During the 1980s and 1990s, the spatial design of most major airline networks underwent a significant change. Networks were spatially centered around one or several "hubs," or around "super-nodes" with many edges. Taking a closer look at the prototypes of spatial network designs, we can see that each has distinct strengths and weaknesses. In theory, three prototypes of spatial network designs prevail (Fig. 87):

- Distributed networks with no centers
- Decentralized networks with pronounced centers
- Centralized networks with only one center

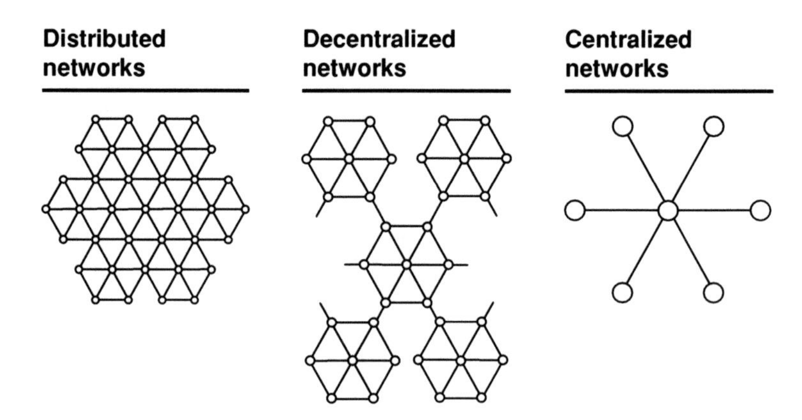

Fig. 87 Spatial network topologies

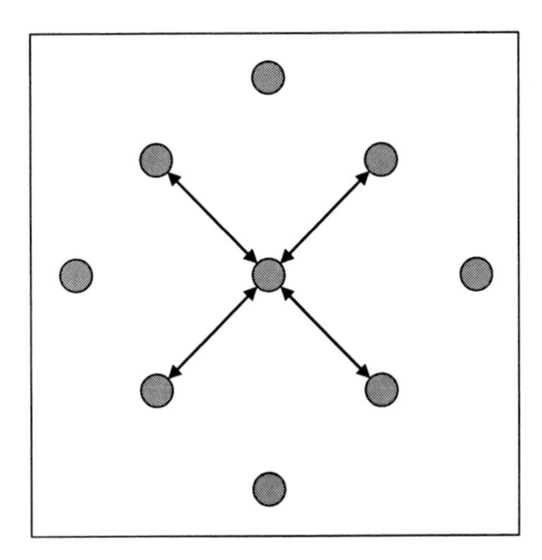

Fig. 88 Paths at the spatial production network layer

Figure 88 imposes a centralized network structure upon the demand structure of the network shown in Fig. 86. In defining this structure and its principal paths, many potential destinations have been excluded from the now defined spatial scope of the network.

A simple yet powerful indicator to assess spatial concentration for airline networks is the ratio of the number of nonstop routes over the number of destinations served within a given network. Burghouwt (2007) reviews various other ways to quantify the degree of spatial concentration within airline networks. He recommends the Gini-Index, developed to quantify income inequalities in social societies, as the most effective way to assess spatial concentration for airline networks. In essence, the Gini-Index compares the actual distribution of seat capacity (allocated to edges) in an airline network with the theoretical distribution if overall seat capacity were randomly distributed among airports (nodes).

10.1.3 The Temporal Layer is Added Next

At the third layer, departure and arrival times are added to the paths of the spatial production network (see Fig. 89). Defining commercially meaningful and operationally feasible departure and arrival times (the purpose) requires in-depth production planning (the means) of all critical resources, such as aircraft, crew, and maintenance docks. Which aircraft will fly which sequence of flights (sequence of edges, or path), and by when? Which crew will fly which sequence of flights? When does a specific aircraft need to return to a specific technical facility for maintenance or repair? Clearly, such questions depend on the structure of the underlying spatial network. The sequence and timing of the path of an aircraft through a network must follow the edges of the underlying spatial production network, and the same applies to the cockpit or cabin crew. The time when an aircraft needs to arrive at a technical facility for regular inspection also depends on the principal spatial production

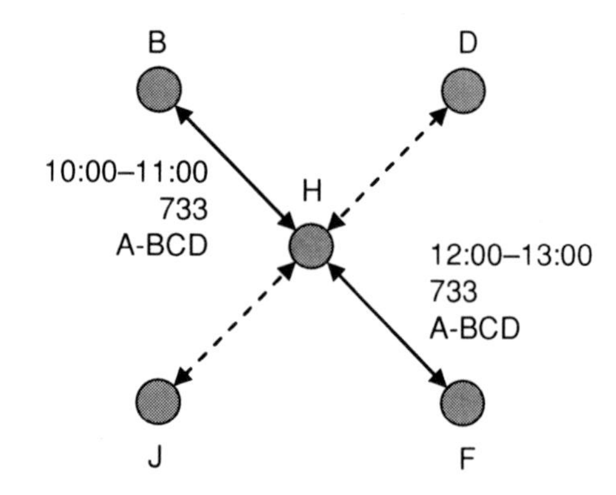

Fig. 89 The temporal network layer. *B* to *F* represents a path at the temporal production layer, connecting two edges (BH and HF) via hub (node) H. The first line of text indicates scheduled departure and arrival times, the second line specifies the respective type of aircraft, and the third line denotes the tail number of the individual aircraft serving these routes

network. Before reliable times can be added to an airline network plan, the type (Boeing 737-300) and tail-number (A-BCD) of a particular aircraft must be defined.

Completion of the planning of the temporal production network layer triggers one important feed-forward and one important feedback loop[7]:

- The publication of the timetable is a **feed-forward** loop from the temporal production to the communication layer. Through publication of the schedule, the airline is committed to deliver accordingly.
- The **feedback** of the rules and constraints of production for optimum planning of spatial network design is a key success factor of LCCs, as the choice of destinations and the spatial design of LCC networks are strongly driven by production considerations.

10.1.3.1 The Pricing Layer is Put on Top of the Production-Driven Layers of an Aviation Network

So long as passengers follow the nonstop or connecting paths through the network as planned by the temporal production layer, these flows would follow the planned paths and not add new ones. Passengers add a new layer, however, when two or more demanded paths compete for the same seat in an aircraft. In Fig. 90, a passenger desires to fly from B to D through hub H, and another passenger demands to fly from B to J through hub H. Assuming a fairly high seat load factor on the flight segment from B to H, both passengers may compete for the same last available seat. As a result, we have two different (and conflicting) flows of potential value. Many highly sophisticated methodologies (revenue management) have been developed to anticipate and resolve such conflicting flows of potential value.

[7] It is tempting to consider the many loops feeding backward from the pricing network layer to the temporal production layer, or the loops feeding forward from temporal production to the communication layer. From a system dynamics point of view, these loops should exert significant oscillations due to such feedback and feed-forward loops. Those oscillations can cause symptoms of organizational conflict.

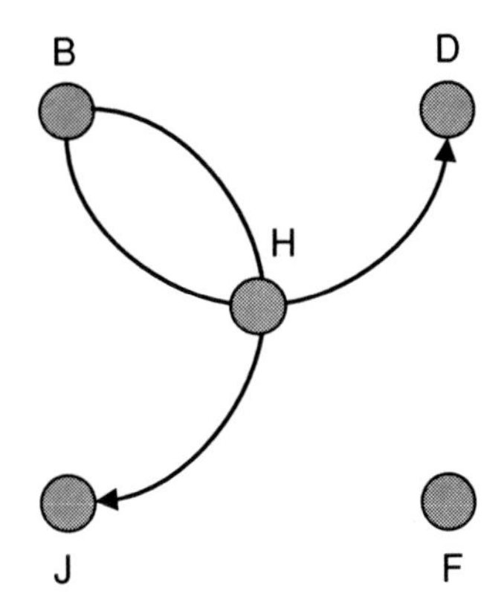

Fig. 90 Paths at the pricing network layer. Two multi-segment paths, *B–H–J* and *B–H–D*, are both competing for the same seat capacity on the segment (edge) *B–H*. Pricing needs to ensure that the highest value demand gets hold of the last remaining seat on *B–H*

10.1.3.2 To Successfully Distribute an Aviation Network, Communication Needs Require a Separate Layer

Airlines need to communicate their network product to the public. They do so by uploading their electronic schedules to the global inventories [global distribution systems (GDS) or computer reservation systems (CRS), such as Amadeus or Sabre]. Airlines also publish their schedules on their respective websites and publicize printed handbook versions of their timetables. At this stage, airlines are not interested in emphasizing which services are connect only and which are nonstop. Instead, they put forward the notion of offering a plethora of city pairs. Figure 91 represents the collection of "services" (connecting every destination with every other destination) of the airline operating the schematic network introduced in Fig. 88. Operational details rank second, though they may be available. This has significant impact on the structural topology of an airline network at this level. For communication and sales purposes, airline managers are highly interested in making their network look as distributed as possible. In the early days of GDSs, this vested interest led to "biased displays" of GDS terminals. The intention was to hide or outright cheat on the production-driven

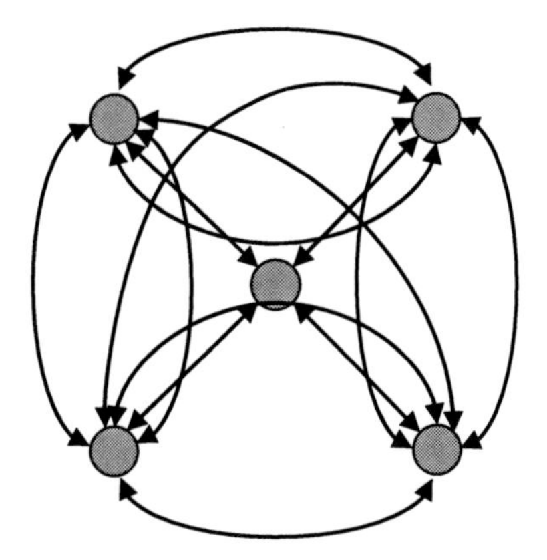

Fig. 91 Paths at the communication network layer. At the communication network layer, airlines pretend to connect with every destination (node)

network elements that deviated from the distribution in the ideal communication schedule.

10.2 How the Concept of Layered Networks can Lead to Better Managed Networks

The four examples below demonstrate the power of the concept of layered networks. Each one illustrates that the striking similarities between the concept of layered networks and the structures in network strategy development, network management IT, and network control are not coincidental.

10.2.1 Better Control of Network Complexity

Complexities bubble up through the hierarchy of layers. A large diversity of fat and thin city pairs at the lowest (demand) network layer will make it more difficult to define a portfolio of comparable destinations (spatial production layer), or to rotate a diverse fleet of aircraft through this layered network. A portfolio of destinations with a large diversity of distances in between them will make it difficult to plan highly efficient aircraft utilization scores. The layer that results when time is added to the schedule (temporal production) is particularly susceptible to complexity.

Network carriers add enormous time complexity through connectivity and corresponding bank structures, while LCCs add a demanding dimension of time complexity by the tight rotational patterns they typically serve. Reducing the complexity at a given layer requires the relaxation of complexity drivers in the layers below. In turn, mastering complexity at a given level permits more complexity at the levels below. The organizational concept of "hub management" (see Sect. 6.5.2 in Chap. 6) illustrates how to address overcomplexity at the higher network layers through a radical segmentation at the second-lowest (spatial) network layer. Then each hub management will manage the complexity of only a fraction of the entire network.

10.2.2 Network Management Key Processes Should Follow the Scope of Individual Network Layers

The hierarchy of the layers from bottom to top—demand, spatial production, temporal production, pricing, and communication—reflect the key processes of network planning (Fig. 71). Mismatching the scope of key planning processes with the scope of these layers is a recipe for organizational failure. This equivalence facilitates the proper allocation of content and respective responsibilities to individual key processes, subprocesses, and organizational units.

There is another important reason why key processes should match network layers. The various layers of networks match the contribution levels of network controlling (see Sect. 6.6 in Chap. 6). Therefore, the network layers, though deduced from a network topology point of view, are equivalent to their respective organizational and accounting processes. This equivalence can greatly simplify the allocation of costs to their "natural" organizational sites.

10.2.3 Integrating Processes—Responsibilities, Accountabilities, Workflows, Decision Criteria, IT systems—at the Same Layer

While the various layers of the network are clearly distinct, that does not mean that only one network is present per layer. Several parallel networks may be found at the operational layers of the network. However, the parallelism of network structures at this layer is artificial, not conceptual. Planners have simply broken down the overwhelming complexity of optimally assigning all kinds of resources at once into parallel and more feasible pieces of work. Aircraft are assigned in one run, cockpit crew in a different run, and cabin crew in a distinct run. These "runs" are the "variants" discussed here. It has recently been shown (Grosche and Rothlauf 2007; Grosche 2009) that genetic algorithms permit the optimization of assigning most resources and constraints (though not yet fully constrained) at once. This serves to overcome the need to subdivide the integrated temporal layer into artificial parallel layer variants. To foster the vertical (across the various layers) consistency of the layered key processes, such horizontal (across a given layer) integration of IT systems is critical. Today, cabin and cockpit crew planning and administration, fleet assignment,[8] rotational planning, tail number assignment,[9] and maintenance dock planning are all distinct cost centers with distinct reporting paths; rarely are they compatible data models and IT tools. The more processes are integrated at a given layer of network management, the smoother they will operate, and the less organizational conflict is likely.

10.2.4 Meta-Layers Advance the Understanding of the Aviation Value Chain

The concept of layered networks can be extended beyond the airline or airport perspective on networks. Air traffic control, for instance, can be conceptualized as a meta-layer of the already layered networks. Contrary to popular belief, the sky is

[8] Fleet assignment: the assignment of a particular type of aircraft to serve a specific rotation.

[9] Tail number assignment: the assignment of an individual aircraft to serve a specific rotation.

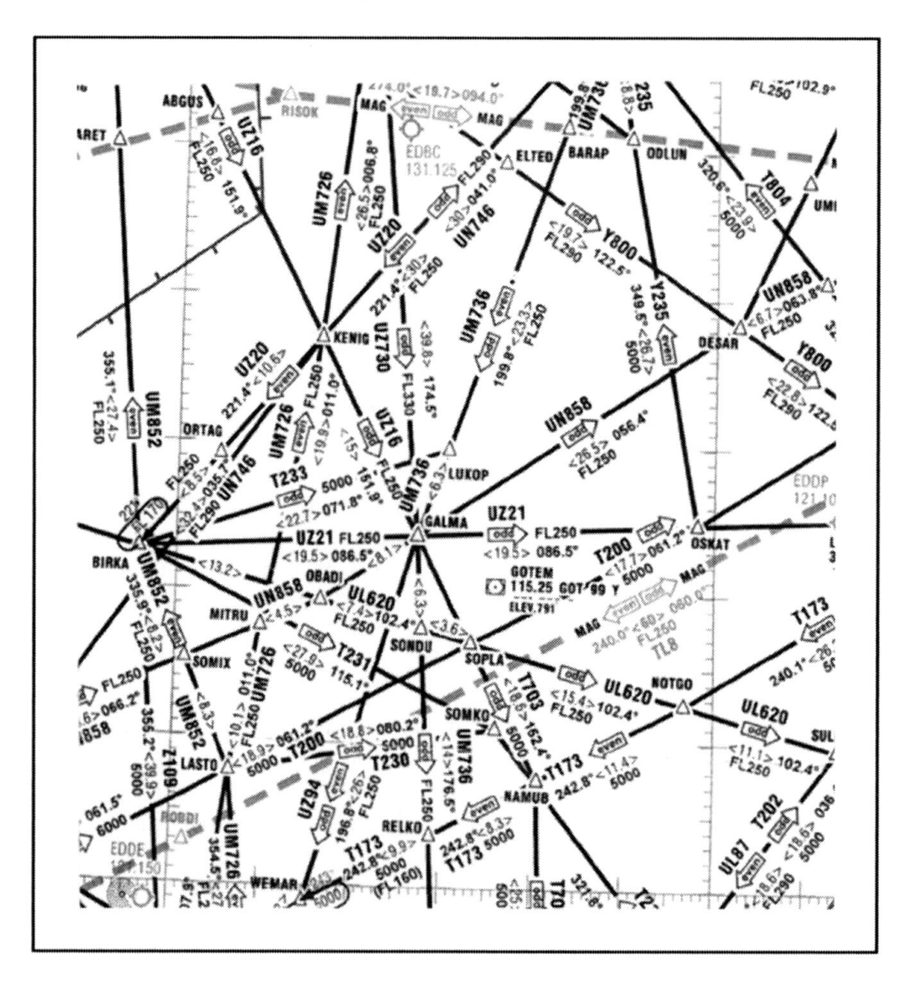

Fig. 92 The meta-layer of aviation networks. A detail of Germany's air traffic navigation map (DFS)

regulated: aircraft may only start after they are given governmental authority to do so. An aircraft may not even push back from its parking position or start its engines without official clearance. Once in the air, pilots must strictly follow the guidance of air traffic controllers at any given time in regard to flight direction and altitude level. The entire sky is tightly packed by a scheme of "streets," flight levels, civil or military zones, and approach or departure areas (Fig. 92). Within this tightly regulated environment called "the sky," air traffic controllers must follow the published timetable, as well as the operational production plan of the "temporal production network." Consequently, all the aircraft movements in the sky, in upper as well as lower airspace, are nothing more than a high layer of the underlying production or product network layers.

References

Bogusch LL (2003) Rethinking the hub-and-spoke airline strategy: An analysis and discussion of American Airlines' decision to depeak its schedule at O'Hare International. MS Thesis, Massachusetts Institute of Technology, Sloan School of Management, Cambridge

Bootsma P (1997) Airline flight schedule development: Analysis and design tools for European hinterland hubs. PhD thesis, University of Twente. Twente

Brandenburger A, Nalebuff B (1996) Co-opetition. Currency Doubleday, New York

Burghouwt G, De Wit J (2005) Temporal configurations of European airline networks. Journal of Air Transport Management 11:185–198

Burghouwt G (2007) Airline network development in Europe and its implications for airport planning. Ashgate, Aldershot

Burghouwt G, Redondi R (2009) Connectivity in air transport networks: Models, measures, and applications. University of Bergamo, Dept. of Economics and Technology Management, Bergamo

Casti JL (1995) Theory of Networks. In: Batten JRD (ed) Networks in Action. Springer, Berlin-Heidelberg

Coldren GM, Koppelmann FS et al (2003) Modeling aggregate air-travel itinerary shares: Logit model development at a major US airline. Journal of Air Transport Management 9(6): 361–369

Danesi A (2006a) Measuring hub time-table co-ordination and connectivity: Definition of new index and application sample of European hubs. European Transport 34:54–74

Danesi A (2006) Spatial concentration, temporal coordination, and profitability of airline hub-and-spoke networks. Ph.D. thesis, Universita di Bologna, Bologna

Dennis N (1994) Airline hub operations in Europe. Journal of Transport Geography 2:219–233

Dennis N (2001) Developments of Hubbing at European Airports. Air & Space Europe (3):51-55

Doganis R, Dennis N (1989) Lessons in Hubbing. Airline Business 3:42–47

Flint P (2002) No Peaking. Air Transport World 39(11):22–27

Goedeking P, Leibold K et al (2008) Schlussfolgerungen. In: Initiative Luftverkehr für Deutschland (ed) Wettbewerbsfähigkeit des Luftverkehrsstandortes Deutschland. Frankfurt am Main

Grosche T (2009) Computational intelligence in integrated airline scheduling. Springer, Berlin-Heidelberg

Grosche T, Rothlauf F (2007) Air Travel Itinerary Market Share Estimation. Working Paper in Information Systems 6, Univ. Mannheim, Mannheim

Handelsblatt (2009) Lufthansa will Billigflieger kopieren. Handelsblatt Nov 17, 2009, Düsseldorf

Holloway S (2003) Straight and Level: Practical Airline Economics, 2nd edn. Ashgate, Aldershot
Jost R (2009) MS Thesis: Statistische Modellierung von Flugplänen. University of Giessen, Giessen
Klein R, Steinhardt C (2008) Revenue Management. Springer, Berlin-Heidelberg
Keypost (2009) Delta and CVG expect more cuts. http://www.kypost.com/content/wcposhared/story/Delta-And-CVG-Expect-More-Cuts/LDf8W5Qe0Em8zXG0tn18kQ.cspx. Accessed April 17, 2010
Lilien GL, Kotler P et al (1992) Marketing Models. Prentice-Hall, Upper Saddle River
MacMenamin S, Palmer J (1988) Strukturierte Systemanalyse. Hanser, München Wien, Prentice Hall, London
Malighetti P, Paleari S et al (2008) Connectivity of the European airport network: "Self-help hubbing" and business implications. Journal of Air Transport Management (14):53-65
Mandel B, Gaudry M et al (1997) A disaggregate Box-Cox Logit mode choice model of intercity passenger travel in Germany and its implications for high-speed rail demand forecasts. The annals of regional science 31:99–120
Maxon T (2010) Southwest adjusts schedule in bid to seek more connecting fliers. The Dallas Morning News, February 21, 2010
Morell P (2007) Airline finance. Ashgate, Aldershot
Paleari S, Redondi R et al (2009) A comparative study of airport connectivity in China, Europe, and the US: Which network provides the best service to passengers?. University of Bergamo, Dept. of Economics and Technology Management, Bergamo
Petroccione L (2007) Delta's Operation Clockwork. Transforming the fundamentals of an airline. Decision Strategies, Inc., Houston
Sala S (2009) Knowledge is Power. Airline Business, September 2009, 72
Schwab K (2009) The Global Competitiveness Report 2009-2010. World Economic Forum, Geneva
Sirower M (1997) The Synergy Trap. The Free Press, New York
Standard Schedules Information Manual (2008) International Air Transport Association. Montreal, Geneva
Vasigh B, Fleming K et al (2008) Introduction to air transport economics. Ashgate, Aldershot
Veldhuis J (1997) The competitive position of airline networks. Journal of Air Transport Management 3(4):181–188

Appendix: Market Research Checklists

Local Markets

- How is demand structured?
  - How many passengers depart from or arrive at the airport (or the airports) within that region or city?
  - What are the important destinations of outbound passengers? From which origins do inbound passengers most frequently depart?
  - What yield levels are or can be achieved (breakdown by distribution channel, destination or origin, and regional submarket)?
- What is the regional expansion of a catchment area? Or, more specifically: Where are the tickets of outbound traffic sold?
  - What growth trends characterize this region? What volatility characterizes this market?
- What seasonal effects must be considered?
- What macroeconomics apply? What infrastructure can be built upon?
  - What are the macroeconomic and political aspects of origins and destinations (GDP, regulatory framework, economic and political stability)?
- What quality and capacity of infrastructure are available? Can airports be expanded? What noise or curfew constraints apply? What air traffic control constraints must be considered? How accessible is(are) the airport(s) at hand? How well is the airport linked with other modes of transportation (car, train)? How efficient are operations?
- What kind of and level of competition has to be considered?
  - With what level of competitive intensity do the airlines have to cope?
- What frequent flyer programs (FFPs) dominate the market? How many active cards are on the market? What level of market clout do these FFPs generate?

  - Who controls which corporate accounts in the market at hand?
- What other airports fight for the same or overlapping catchment area?

City Pairs

- How is demand structured?
  - How many passengers want to fly from A to B?
  - What are the time-of-day and day-of-week preference profiles of passengers for this particular O&D?
  - What price elasticity patterns apply for which passenger segment?
  - What growth trends characterize this O&D? How volatile is the growth?
  - What seasonal effects must be considered?
- What quality of connections is offered?
  - Is the city pair served point-to-point or transfer only?
  - How awkward or comfortable is transfer? How many and how attractive are the existing options to connect?
- What macroeconomics apply? What infrastructure can be built upon?
  - What are the macroeconomic and political aspects of origins, the connecting point, and destinations (GDP, economic and political stability)?
  - What quality and capacity of infrastructure are available at the origin, connecting point, and at the destination? Can airports be expanded? What noise or curfew constraints apply? What air traffic control constraints must be considered? How accessible is(are) the airport(s) at hand? How well is the airport linked with other modes of transportation (car, train)? How efficient are operations?
- What infrastructure conditions must be considered?
  - Runway, terminal, apron, land-side capacity, and infrastructure?
  - Minimum connecting times? Curfews?
  - Punctuality and lost baggage statistics?
- What kind of and level of competition has to be considered?
  - With what level of competitive intensity do the airlines have to cope?
  - What other airports are fighting for the same transfer traffic? What are their or "my" strengths and weaknesses to attract transfer traffic to the particular O&D at hand?

Painting and Pricing

(By David Tait for TravelIndustryToday.com)

Buying paint from a hardware store ···

Customer: Hi, how much is your paint?
Clerk: We have regular quality for $12 a gallon and premium for $18. How many gallons would you like?
Customer: Five gallons of regular quality, please.
Clerk: Great. That will be $60 plus tax.

Buying paint from an airline ···

Customer: Hi, how much is your paint?
Clerk: Well, sir, that all depends.
Customer: Depends on what?
Clerk: Actually, a lot of things.
Customer: How about giving me an average price?
Clerk: Wow, that's too hard a question. The lowest price is $9 a gallon, and we have 150 different prices up to $200 a gallon.
Customer: What's the difference in the paint?
Clerk: Oh, there isn't any difference; it's all the same paint.
Customer: Well, then, I'd like some of that $9 paint.
Clerk: Well, first I need to ask you a few questions. When do you intend to use it?
Customer: I want to paint tomorrow, on my day off.
Clerk: Sir, the paint for tomorrow is $200 paint.
Customer: What? When would I have to paint in order to get $9 paint?
Clerk: That would be in three weeks, but you will also have to agree to start painting before Friday of that week and continue painting until at least Sunday.
Customer: You've got to be kidding!
Clerk: Sir, we don't kid around here. Of course, I'll have to check to see if we have any of that paint available before I can sell it to you.
Customer: What do you mean check to see if you can sell it to me? You have shelves full of that stuff; I can see it right there.
Clerk: Just because you can see it doesn't mean that we have it. It may be the same paint, but we sell only a certain number of gallons on any given weekend. Oh, and by the way, the price just went up to $12.
Customer: You mean the price went up while we were talking?
Clerk: Yes, sir. You see, we change prices and rules thousands of times a day. And since you haven't actually walked out of the store with your paint yet, we just decided to change. Unless you want the same thing

to happen again, I would suggest you get on with your purchase. How many gallons do you want?

Customer: I don't know exactly ··· maybe five gallons. Maybe I should buy six gallons just to make sure I have enough.

Clerk: Oh, no sir, you can't do that. If you buy the paint and then don't use it, you will be liable for penalties and possible confiscation of the paint you already have.

Customer: What?

Clerk: That's right. We can sell you enough paint to do your kitchen, bathroom, hall, and north bedroom; but if you stop painting before you do the other bedroom, you will be in violation of our tariffs.

Customer: But what does it matter to you whether I use all of the paint? I already paid you for it!

Clerk: Sir, there's no point in getting upset; that's just the way it is. We make plans based upon the idea that you will use all of the paint; and when you don't, it just causes us all sorts of problems.

Customer: This is crazy! I suppose something terrible will happen if I don't keep painting until Sunday night?

Clerk: Yes, sir, it will.

Customer: Well, that does it! I am going somewhere else to buy paint!

Clerk: That won't do you any good, sir. We all have the same rules. You might as well just buy it here, while the price is still $13.50. Thanks for flying—mean painting—with our airline.

Index

Breinigsville, PA USA
17 February 2011

255750BV00006B/2/P